The Latin American Studies Book Series

Series editors

Eustógio Wanderley Correia Dantas, Fortaleza, Brazil
Jorge Rabassa, Ushuaia, Argentina
Andrew Sluyter, Baton Rouge, USA

The Latin American Studies Book Series promotes quality scientific research focusing on Latin American countries. The series accepts disciplinary and interdisciplinary titles related to geographical, environmental, cultural, economic, political and urban research dedicated to Latin America. The series publishes comprehensive monographs, edited volumes and textbooks refereed by a region or country expert specialized in Latin American studies.

The series aims to raise the profile of Latin American studies, showcasing important works developed focusing on the region. It is aimed at researchers, students, and everyone interested in Latin American topics.

More information about this series at http://www.springer.com/series/15104

Rogelio Daniel Acevedo
Jesús Martínez Frías
Editors

Geoethics In Latin America

Springer

Editors
Rogelio Daniel Acevedo
Centro Austral de Investigaciones Científicas (CADIC-CONICET)
Ushuaia, Tierra de Fuego
Argentina

Jesús Martínez Frías
Facultad de Ciencias Geológicas
Instituto de Geociencias IGEO (CSIC-UCM)
Madrid
Spain

ISSN 2366-3421 ISSN 2366-343X (electronic)
The Latin American Studies Book Series
ISBN 978-3-030-09216-0 ISBN 978-3-319-75373-7 (eBook)
https://doi.org/10.1007/978-3-319-75373-7

Printed on acid-free paper

This Springer imprint is published by the registered company Springer International Publishing AG part of Springer Nature
The registered company address is: Gewerbestrasse 11, 6330 Cham, Switzerland

Foreword

Ethics (morals), starting from Greek antic times, was understood as a measure of dominance of human over himself, the indicator of how a person is responsible for himself, for what he does. Geoethics that developed from ethics, and based on its forms of comprehension of general methods of relations of humans to the world, to be understood, primarily, as the responsibility of a human for his actions in relation to the inanimate nature.

To my mind, in the triad of "culture, morality and economy" morality indubitably takes the first position. It represents the ground zero. All other aspects of human life, their successes in economic and cultural development, their place in the world are determined by those moral attitudes, which are not only proclaimed but also firmly adhered to in the society and at all its levels.

In our contemporary world, a special significance is given to general philosophic understanding of the evolving social–economic situation, interrelations between people and entire nations, interrelations of humans with the natural environment, modern views about the natural sue issues, including use of mineral resources and other sites of inanimate nature.

A critical analysis of value orientations, deeper understanding not only of economic but also of moral basics of use of sites of inanimate nature is underway. Combined analysis of economic, legal and moral norms is getting greater importance in development of concepts of long-term development of mineral-raw complexes of entire countries, development of strategic approaches to complex use of raw materials and fair distribution of profits earned, and experts' review of large scale mineral resource use project.

As one of the applied branches of philosophy in the set of geological sciences, geoethics takes a special position. Studying the possibility of use of ethical principles and norms applicable to the activities in the industry of resource supply, resource use and resource consumption, it brings together practically all fields of geological activities into one point and can be considered as one of the most important geological sciences.

Development of geoethics and its results were officially acknowledged at the XXX International Geological Congress held in Beijing in 1996. It was in Beijing that for the first time in practice of international geological congresses, a subsection on the issues of geoethics was established within the geological education section.

During those days, the notion of "geoethics" was defined in very narrow terms—as a science about ethical principles and norms while using mineral resources—and was based only on essence features of mineral resources (exhaustibility, non-renewability, geographical unevenness of their distribution, belonging not only to existing but also to future generations). By occurrence of new era challenges (boom of non-renewable natural resource use, economic, social–cultural and political globalisation, the necessity in implementation of a strategy of sustainable future, ecological crisis threat, planetary climate changes, terrorism) and significant expansion of the subjects and objects of study and the definition of the term has changed. Now, geoethics is considered as a science, which studies moral basics of relations in the system of "humans-abiotic nature of the planet Earth and other celestial bodies". By proposing new conceptual approaches of mutual relations in the "humans-abiotic nature", geoethics is undertaking a real attempt to create a general planetary ethics, which is based on such key factors like justice and responsibility.

Currently due to expansion of disciplines and object of study, development of Geoethics has entered into a stage of differentiation of accumulated knowledge and formation of main scientific orientations: general theoretical, methodological, applied, economic, social, ecological, educational, cultural and moral-religious orientations.

This book "Geoethics in South America" will familiarise the readers with specific geoethical issues and dilemmas that are characteristic to this continent.

Nataliya Nikitina
First IAGETH Vice-President

Contents

Part I
Introduction

Chapter 1
Geoethics

Rogelio Daniel Acevedo and Jesús Martínez Frías

Abstract Geoethics was born in 1991 at the junction of ethics and geology. Dr. Vaclav Nemec is considered the father of this discipline. Geoethics has been accepted by both Earth and Social Sciences because of the necessity of an appropriate ethical attitude to the whole geosphere and of a critical analysis of geoethical dilemmas and finding ways how to solve them.

Keywords Geoethics · Geology · Ethics

1.1 Introduction

Geoethics is a key discipline in the field of Earth and Planetary Sciences, which includes scientific, technological, methodological and social–cultural aspects (e.g. sustainability, development, museology), but also the necessity of considering appropriate protocols, scientific integrity issues and a code of good practice, regarding the study of the abiotic world (Fig. 1.1). Geoparks and Geosites are excellent initiatives too, not only for the Earth, but also for the Moon, Mars and beyond, which are narrowly related to the ethical approach.

Geoethics and geodiversity have gained importance in recent decades in the context of geosciences and their interdisciplinary links. Geoethics was formerly promoted in 1991 as a new discipline in the framework of Earth Sciences, linking ethics and geology, and involving scientific and societal aspects from theoretical and practical approaches. Geodiversity was used for the first time in 1993 as the

R. D. Acevedo (✉)
CADIC-CONICET, Centro Austral de Investigaciones Científicas,
Ushuaia, Tierra de Fuego, Argentina
e-mail: acevedo@cadic-conicet.gob.ar

J. M. Frías
Facultad de Ciencias Geológicas, Instituto de Geociencias IGEO
(CSIC-UCM), Madrid, Spain
e-mail: j.m.frias@igeo.ucm-csic.es

R. D. Acevedo and J. M. Frías (eds.), *Geoethics In Latin America*,
The Latin American Studies Book Series,
https://doi.org/10.1007/978-3-319-75373-7_1

Fig. 1.1 Field geologists. Mirny, Siberia, Russia

geological equivalent of biodiversity. It has had considerable conceptual development and noteworthy institutional support and now warrants the status of a geological paradigm, incorporating in its definition all the variety of rocks, minerals and landforms and the processes which have formed these features throughout geological time. The incorporation of the geoethical and geodiversity issues in planetary geology and astrobiology studies would enrich their methodological and conceptual character.

Geoethics has applications to, among others, the research of meteoroids and meteorites, the geological study of terrestrial analogs, the correct use of mineralogical and geochemical analytical procedures, the interrelation with bioethics and organics versus inorganic contamination in Planetary Protection and many other aspects associated with scientific integrity in the context of planetary geology and astrobiology. Likewise, geodiversity, following the recently defined and approved IUCN guidelines and recommendations, would have an unambiguous use in Space as Planetary Geodiversity, valuing and safeguarding the different types and groups of meteorites as representatives of the planetary bodies of our solar system, as well as considering the worth of the abiotic nature of the planets in itself.

1.2 Overall Objectives

Geoethics is an interdisciplinary field between geosciences and ethics which involves Earth and Planetary Sciences as well as applied ethics. It deals with the way of human thinking and acting in relation to the significance of the Earth as a system and as a model. Not only geoeducational, scientific, technological, methodological and social–cultural aspects are included (e.g. sustainability, development, geodiversity and geoheritage, prudent consumption of mineral resources, appropriate measures for predictability and mitigation of natural hazards, geosciences communication, museology), but also the necessity of considering appropriate protocols, scientific integrity issues and a code of good practice, regarding the study of abiotic world. Studies on planetary geology (sensu lato) and astrobiology also require a geoethical approach.

The main objectives are:

- To promote and encourage the advancement of geoethics—primarily through educational and publishing activities and research;
- To foster geoethical ways of thinking and acting—especially in geosciences (because of their significance for any support and help in developing realistic sustainability concepts);
- To improve teaching and training in geoethics;
- To realize the goals defined by the International Declaration of Geoethics approved in October 2011 by the AGID Working Group for Geoethics; and
- To collect, evaluate and disseminate the results of geoethical activities on a worldwide basis.

One of the most important messages of geoethics to the population of our planet concerns the fact that the future of the Earth is determined not only by anthropogenic influences but also by long-term exogenous and endogenous natural processes, often accompanied by natural disasters beyond human control. The most significant of these events are often beyond the reach of human memory. Only Earth and Space Sciences are able to serve as mediators for any research needed in detecting the behaviour and predictability of such phenomena. In the light of this knowledge, it seems necessary to modify the mostly oversimplified ideas about the environment sustainability, by an appropriate geoeducation and by a geoethical approach. It is necessary to realize that the so-called abiotic nature has its own dynamic evolution and that its regularities and laws need to be known and understood in order to improve any forecasting and mitigation of important catastrophes and climate changes.

1.3 Relate Goals to Overall IUGS Scientific Objectives

The IUGS aims to promote development of the Earth sciences through the support of broad-based scientific studies relevant to the entire Earth system; to apply the results of these and other studies to preserving Earth's natural environment, using all natural resources wisely and improving the prosperity of nations and the quality of human life; and to strengthen public awareness of geology and advance geological education in the widest sense.

The International Association for Geoethics (Fig. 1.2) was approved in 2013 as an affiliated organization to the IUGS. Knowing the significance of the link between ethics and geosciences and the different views and approaches regarding the diversity of the Earth's cultures, IAGETH aims to become a key tool of IUGS for developing geoethical activities realizing about the significance of people and society. The main goals of IAGETH follow the IUGS scientific guidelines, in accordance with its main role for fostering geoethics in connection with geoeducation and with the academic, scientific and professional duties of geoscientists, stressing the awareness for the benefit of society.

Fig. 1.2 Emblem of the International Association for Geoethics (IAGETH) *Credit* IAGETH

Chapter 2
Geoethics: Basic Concepts and Its Potential for UNESCO Geoparks

Joan Poch

Abstract We only have one home in space, the Earth, and its habitability depends on the planet as a system, formed by the geosphere, hydrosphere, atmosphere and biosphere, continued with its natural dynamics. We humans multiply quickly and we occupy more and more space and need more and more natural resources. Thus, in the field of Earth Sciences, our economic activities interfere with the operation of this system on an unprecedented scale. Scientists talk about a new geological age (the Anthropocene) to show that our species is leaving as many traces in the geological record as a geological force. For our own survival, it is necessary to move quickly towards a more sustainable relationship with the Earth and its limited resources. With respect to the geosphere, geologists have a responsibility to drive this change. Geoethics can provide guidelines and tools so that the behaviour and professional actions of geologists may facilitate this path towards sustainability and protection of the values of the geosphere. In order to communicate these good ethical practices to society, UNESCO Global Geoparks are suitable spaces, due to its model of sustainable management of geosphere values for the benefit of local communities.

Keywords Geoethics · Earth system · Anthropocene · Global change
UNESCO Global Geoparks

2.1 Introduction

In a context of globalization, where world population and demand for natural resources are rapidly increasing, mankind has the potential to increasingly influence the functioning of the Earth system in a manner comparable to geological forces,

---

J. Poch (✉)
Departamento de Geología, Universitat Autònoma de Barcelona, (Despacho C2/130).
Calle de La Vall Moronta, 08193 Cerdanyola del Vallès, Barcelona, Spain
e-mail: Joan.Poch@uab.cat

R. D. Acevedo and J. M. Frías (eds.), *Geoethics In Latin America*,
The Latin American Studies Book Series,
https://doi.org/10.1007/978-3-319-75373-7_2

an idea anticipated by Vladimir Vernadsky, a renowned Russian–Ukrainian mineralogist (1863–1945). He also introduced the concept of noosphere or the sphere of human cognizance and the theory of interdependence of geosphere, biosphere and noosphere.

In the field of Earth Sciences, problems arising from the interaction between human activities and nature are complex. In general, the solutions are also complex and they carry an ethical dimension that forces us to consider our responsibility in these activities (Peppoloni and Di Capua 2015).

While some traditional societies try to live in a sustainable way and are well aware of the limitations of our planet's natural resources (Fig. 2.1), the most technologically developed societies consume resources at such a speed that it seems they do not think too much about the needs of future generations.

Fig. 2.1 Our planet seen from space 35,000 km (22,000 miles) above the Earth. This type of imagery helps us to be more aware of the finitude of our home and its resources. *Source* NASA/GSFC/Reto Stöckli, Nazmi El Saleous and Marit Jentoft-Nilsen (2000). *Credit NASA/GSFC/Reto Stöckli, Nazmi El Saleous and Marit Jentoft-Nilsen*

This chapter will deal with how geoethics provides guidance to Earth Science professionals to better target their actions towards achieving a more sustainable relationship of humanity with the planet.

In the first place, the concept of geoethics is defined as an emerging scientific discipline. Next, the concepts of "Earth System", "Gaia", "Anthropocene", "Global Anthropogenic Change" and "Earth System goods and services" are outlined, delimiting the framework of geoethics. Next, a synthesis of the values of the geosphere that are recognized by geoethics is presented. The principles and activities of geoethics are described below. The following section refers to the Ethical Code of the UNESCO Global Geoparks and outlines examples of geoethics values that can be integrated into the usual activities of the Geoparks. The last section is reserved for conclusions.

2.2 What Is Geoethics and Why Is It Necessary?

The word "geoethics", as used starting from the early 1990s (Savolainen 1992), signifies the duty of mankind to behave responsibly and become the natural consciousness of the planet. According to Almeida and Vasconcelos (2015), geoethics deals with the ethical, social and cultural implications of the behaviour and professional activities of geologists.

Given the limitation of natural resources and the fragility of the planet Earth, it is considered that: "Geoethics is concerned with promoting the analysis of the use of our natural resources, to promote correct information on natural hazards and the development of technologies that are respectful with the environment, while extending its principles for planetary protection" (Martínez-Frías et al. 2011).

A more complete definition: "Geoethics consists of research and reflection on the values which underpin appropriate behaviours and practices, wherever human activities interact with the geosphere. Geoethics deals with the ethical, social and cultural implications of Earth Sciences education, research and practice, providing a point of intersection for Geosciences, Sociology, Philosophy and Economy. Geoethics represents an opportunity for Geoscientists to become more conscious of their social role and responsibilities in conducting their activity. Geoethics is a tool to influence the awareness of society regarding problems related to geo-resources and geo-environment" (International Association for Promoting Geoethics—IAPG).

In the quest for environmental, economic and social sustainability, the need for geoethics arises from the social demand for useful tools to prevent or manage the increasing impact of human activities on the geosphere, the mineral part of the Earth system.

Now that we are capable of destroying the planet, geoethics wonders how to act to ensure our survival and that the Earth continues to function as a system. It talks about changing our habits of life and understanding our relationship with the planet. Geoethics is expected to provide us with guidelines and tools for this.

When the first monograph on Geoética was presented at the 34th International Geological Congress (Brisbane, Australia, 2012)—"Geoethics: theory, principles, problems" (Nikitina 2016)—there were just very few scientists in the world, not more than 50 in total, researching geoethics. During the last years, national geoethical societies have been established in many countries. On an international level, these societies have been combined with two scientific associations: (a) International Association for Geoethics—IAGETH (http://www.icog.es/iageth/) and (b) International Association for Promoting Geoethics—IAPG (http://www.geoethics.org/).

2.3 Geosphere and Earth System: Understanding Planetary Dynamics

The Cartesian scientific method, which consists of addressing the problems of reality by dividing it into more affordable parts (reductionism), is often encountered with difficulties in its application when the object of study is essentially complex, as in the case of planet Earth.

In the middle of the twentieth century, the Austrian biologist and philosopher Ludwig von Bertalanffy formally updated the idea that the constitutive characteristics cannot be explained from the description of the isolated parts. An idea that refers to the well-known phrase "the whole is more than the sum of the parts". According to Pascual (2011), the essence of Bertalanffy's contribution is the existence in the "whole" of new or "emerging" characteristics that are not detected in their isolated parts. Thus, when a complex unit is decomposed, those emerging properties are no longer manifested and cannot be studied.

In order to overcome these limitations, Bertalanffy proposed a general system theory (Bertalanffy 1968), in which "the system" is defined as a complex set of interacting elements, emphasizing the importance of interactions over the specific characteristics of each element.

In the Earth Sciences, the systemic approach is evidenced by the transdisciplinary methodology involved in the study of the water cycle or global tectonics, to cite two examples in which it is necessary to understand the interaction between the geosphere, the atmosphere, the hydrosphere and the biosphere, as subsystems of the Earth system, in addition to the noosphere. In these cases, it is sought to understand a reality constituted by "the usually synergic (and therefore not coincidental with the sum) interaction of the content of each of the different disciplinary areas of study" (Pascual 2011).

Analysing the self-regulating physiological processes of planet Earth, the independent British scientist James Lovelock, with the support of the American microbiologist Lynn Margulis, took the concept of the Earth system further and began formulating the "Gaia hypothesis" in the early 1960s. With this name, which evokes the goddess of nature in various cultures, such as Gea, Pachamama or Mother Earth, Lovelock refers to the Earth as a "living planetary entity", without necessarily implying the existence of consciousness. Lovelock considers that living matter has evolved in conjunction with the geosphere and, therefore, life would be a non-individual planetary property (Lovelock 1979).

Although Gaia's hypothesis has detractors, the idea that life contributes to maintaining habitability on Earth remains an acceptable scientific idea, providing an explanation of the behaviour of the Earth system as a consequence of natural and anthropic imbalances. In addition, visualizing our planet as a "superorganism" facilitates to many people the consideration of the ethical issues that concern our relationship with the Earth.

2.4 Anthropocene and Anthropogenic Global Change

In 2000, the term Anthropocene was coined by Crutzen and Stoermer to describe the present geological epoch, in which human activity dominates many of the processes acting on the surface of the planet. The expression has been widely adopted (Crutzen 2002; Ruddiman 2003), and its status is currently being assessed by the Anthropocene Working Group of the Subcommission on Quaternary Stratigraphy (International Commission on Stratigraphy). Meanwhile, its usage remains informal.

The notion of the Anthropocene can be useful in quantifying alterations in the natural evolution of the Earth system that result from the evolution of human history (Monastersky 2015). It also shows that, in order to understand global processes, a transdisciplinary synthesis of the natural sciences, the social sciences and the humanities is required (Bohle 2014).

Humanity is an engineering species. Biological evolution has been paralleled by the ability to manufacture tools, and from the outset, the evolution of mankind involves the modification of the environment to obtain resources (Smith and Zeder 2013). Humans are altering the Earth, and anthropic activities have left a mark on the geological record since the establishment of agriculture in the Neolithic (Foley et al. 2013). Since the Industrial Revolution, economic activities to obtain resources have left traces that are increasingly recognized globally (Ellis et al. 2010).

The concept of engineering species has a philosophical side: the technological transformation of the environment is about how people manage the appropriation of natural resources according to their chosen system of values, culture and way of life. Such a choice should carry an ethical responsibility.

Global anthropogenic change is a historical process. It is the result of how people interact with the Earth system, especially with the geosphere and biosphere, in order to support a rapidly growing population of several billion people. Thus, this global change refers to the patterns of production and consumption of goods that cause energy flows and that modify the dynamics of the Earth system on an unprecedented scale.

Managing these changes—environmental impacts—includes actions such as "prevention", "adaptation to change" or "mitigation" but have never before been carried out on a planetary scale (Palsson et al. 2012). In short, this management is the main challenge and entails ethical issues, which are related to the responsibility of the professional practice of geoscientists related to these changes. This professional ethics can be synthesized as the set of principles that define the rights and responsibilities of scientists in their relations with other scientists, the media and other people working in land planning and management (Chalk et al. 1980).

Maintaining integrity and ethics in today's competitive world is a difficult challenge. The scientific results and the researchers themselves are under intense scrutiny, often mediated. Sometimes the effectiveness of science or the scientists themselves is attacked. In addition, many subjective opinions are mixed, reflecting different worldviews, and in many cases, the cause–effect relationships between human actions and the geosphere are difficult to determine.

2.5 Earth System Goods and Services

To know more precisely what the conflicts arising from the interaction of anthropic activities with the natural dynamics of the Earth system may be, we consider below the goods and services that we obtain from the geosphere.

The challenges of peak oil, peak phosphorus (where the demand for oil and phosphorus may soon outstrip supply (Cordell et al. 2009) and climate change demonstrate the existence of limits to the rate at which humanity can consume the planet's geophysical resources.

The so-called Earth System goods and services (Steffen et al. 2011) could be classified according to the ecosystem services categorization of the Millennium Ecosystem Assessment (MA 2005) in three basic types:

Provisioning goods and services: commonly known as "natural resources", these include food, fibre, fresh water, fossil fuels, phosphorus, metals and other materials derived from Earth's geological resources.

Supporting services: they are also sometimes called "environmental resources". These include nutrient cycling, soil formation and primary production. All are necessary to support, for example, well-functioning agricultural systems.

Regulating services: these are generally considered as "free services" provided by nature. Two of the most well known of these are the ecological control of pests and diseases and regulation of the climate system through the uptake and storage of carbon by ecosystems.

2.6 Geosphere Values

To develop ethical conduct in the management and preservation of the geosphere, we must consider the value we can give to its elements (geodiversity), including geological systems from the global scale (Earth) to the elemental scale (atoms and molecules). According to several authors (Sharples 1995; Eberhard 1997; Prosser 2002; Gray 2004; Nikitina 2012), seven types of values can be defined:

Intrinsic value: this value goes back to ethical approval that some things (in our case, geodiversity) have a value simply because they exist, not just because they can be used by people (utilitarian value). It is the most difficult value to understand, since it includes ethical and philosophical dimensions of the relations between society and nature.

Cultural value: the value placed by society on some aspect of the geodiversity by reason of its social or community significance.

Aesthetic value: it refers to the visual appeal (and those of other senses) provided by the physical environment.

Economic value: it refers to put a financial value on all environmental assets.

Functional value: it refers to the functional role in environmental systems.

Research and educational value: provides the starting point for scientific work. The geological record needs to be conserved for future studies.

Information value: the value of the information of the abiotic element (by analogy with the genetic diversity of the species). Each geological object has its own unique geological information, and the complexity of this information is directly related to the place it occupies in the Earth system.

Taking into account these values, Nikitina (2012) presents seven geoethic principles for geoconservation:

1. Do not intervene in geological processes on a large scale. They are only allowed on a local scale when necessary to preserve human life.
2. People do not have the right to reduce geodiversity except to meet basic needs.
3. The recognition of the need to establish effective limits of consumption and use of mineral resources, as components of the geological environment, based on the stabilization of the world population. Replace the linear degradation of mineral resources by promoting the use of renewable resources and recycling of mining waste.
4. Opposition to the deregulated market economy, especially if it is a mineral resource-based economy.
5. The use of mineral resources, objects and elements of the geological environment should be based on the recognition of the objective laws of development and the interactions between the geosphere and society and on the economic interests of society, guaranteeing the rights of Citizens to benefit from the use of the subsoil.
6. Geological heritage preservation policies and strategies should be developed as a complex interactive system of institutions and individuals: governments, public and social organizations, at the global, national and regional levels, along with geoscientists, and “heritage consumers” (e.g. tourists of UNESCO Geoparks and geological museums).
7. In the complex system “humanity and geosphere”, the whole is more important than the part, so that the use of a geological object is allowed if it does not lead to the global reduction of geodiversity.

2.7 Behave and Act Ethically: Areas of Application of Geoethics

In analogy with the role of ethics as a teaching of morality and the moral basis of social relations, geoethics is a tool for confronting problems and dilemmas in anthropic actions that have an impact on the geosphere. It includes a wide range of professional activities of geoscientists, such as: research, communication, education or natural risk management. Changes in the dynamics of Earth systems can be made, when necessary, through responsible behaviours and proper practices towards geodiversity and biodiversity (Ellis and Haff 2009).

According to Peppoloni and Di Capua (2012), "Geoethics focuses on some of the most important environmental emergencies: it encourages a critical analysis of the use of natural resources, promotes careful management of natural risks and fosters the proper dissemination of the results of scientific studies, including the development of environmentally friendly technologies".

Moreover, geoethics fosters the development of geoparks (Eder 2004; Zouros 2004; McKeever 2013) and geotourism (Newsome and Dowling 2010; Dowling 2011).

As a synthesis of the issues that should be dealt with geoethics is quoted below (box 1), the Final Statement of the 35th IGC—International Geological Congress in Cape Town, South Africa (2016).

Box 1: Cape Town Statement On Geoethics (2006) "It is essential to enrich the roles and responsibilities of geoscientists towards communities and the environments in which they dwell, as well as paying attention to each scientist's individual conscience and relationships with colleagues. Human communities will face great environmental challenges in the future".

Geoscientists have know-how that is essential to orientate societies towards more sustainable practices in our conscious interactions with the Earth system. Applying a wider knowledge-base than natural sciences, geoscientists need to take multidisciplinary approaches to economic and environmental problems, embracing (geo)ethical and social perspectives. Geoscientists are primarily at the service of society. This is the deeper purpose of their activity.

In the coming years, especially when addressing matters like energy supply, use of geo-resources, land management, pollution abatement, mitigation of geo-risks, and climate change adaptation and mitigation, ethical and social issues will be central in scientific discussion and in public debate. In addition, handling large quantities of data, science and risk communication, education strategies, issues of research integrity, anti-harassment and anti-discrimination policies, gender balance and inclusion of those living with disabilities will be major topics for geoscientists.

Raising the (geo)ethical awareness and competences of the members of the geoscience community is essential, also to increase trust and credibility among the public. This can best be achieved in the near future by two means: by promoting more effectively existing guidance such as codes of ethics/conduct and research integrity statements; and by introducing geoethics into geoscience curricula, to make geoethics a basic feature of the training and professional activity of geoscientists."

Source: IAPG Executive Council (October 2016).

2.8 Geoethics and UNESCO Global Geoparks

Although the UNESCO Global Geoparks Network (GGN) has not yet formalized its membership in any Geoethics Association, it has a Code of Ethics for UNESCO Global Geoparks and Global Geopark Professionals.

The GGN Code of Ethics has been prepared by the Executive Board of the Global Geoparks Network. The GGN Code of Ethics for GGN Members was endorsed at the 20th Meeting of the GGN Executive Board on 12 August 2016, and will be finally approved by the 1st Ordinary General Assembly of GGN, on 28 September 2016. The GGN Executive Board is the decision-making body on any issue dealing with the implementation of the GGN Code of Ethics.

The Code reflects principles generally accepted by the UNESCO Global Geoparks community. It provides a means of professional self-regulation and sets minimum standards of conduct and performance to which all GGN Members throughout the world may reasonably aspire as well as providing a statement of reasonable community expectation from the Geopark profession. The GGN Code of Ethics is organized into seven principles (box 2).

Box 2: Principles of The UNESCO Global Geoparks Network Code of Ethics

1. GGN Institutional Members Administration and Management: GGN Institutional Members, as UNESCO Global Geoparks, should contribute (subject to national legislation) to the protection and rational management of the geological heritage sites as well as to other tangible and intangible natural and cultural heritages within their territory. Through the promotion of these different expressions of heritage they should provide sustainable economic development for the direct benefit of the areas communities.
2. GGN Members and Geological Heritage: GGN Institutional Members have the duty to protect, preserve and promote their Geological Heritage. GGN Institutional Members must respect national, regional, local and

indigenous laws relating to the protection of geological heritages. The defining geological heritage sites within a UNESCO Global Geopark must be legally protected.

3. GGN Members and Natural and Cultural Heritage: GGN Institutional Members need to link their geological heritage with all other aspect of their territories heritages. They have a specific responsibility to assist and support the protection, conservation and promotion of their natural, cultural and intangible heritages and biodiversity.
4. GGN Institutional Members and Intangible Heritage: Intangible Heritage is, in all the world, the more endangered heritage. This specific heritage linked with the intima relation between man and the Earth and is often related with the Geological heritage of the places. The GGN Institutional Member management body has to be involved in the conservation, knowledge, transmission and promotion of this heritage.
5. GGN Institutional Members supporting sustainable development: GGN Institutional Members need to assist an integrated and sustainable economic development for its territory. A GGN Institutional Member has direct impact on the territory by influencing its inhabitants' living conditions and environment. The objective is to enable the inhabitants to re-appropriate the values of the territory's heritage and actively participate in the territory's cultural revitalization as a whole. The territorial sustainable economic development strategy should be in accordance with the principals defined in the 2030 United Nation Agenda for Sustainable Development. The sustainable development of a GGN Institutional Member is only possible with the full participation, of all its population and stakeholder without any kind of discrimination of race, colour, gender, language, religion, political or other opinion, national or social origin, prosperity, birth, sexual orientation or other status.
6. GGN Members must operate in a professional manner: Employees, partners and associates of the GGN Institutional Members and GGN Individual Members should behave in a way that respects the dignity and philosophy of the GGN. They should safeguard the public against illegal or unethical professional conduct. Every opportunity should be used to inform and educate the public about the aims, purposes, and aspirations of the GGN to develop a better public understanding of the contributions of the GGN to society.
7. A standard of conduct which applies to all GGN Institutional members employees, to the Global Geopark professionals and volunteers: Geoparks aim to enable every man, woman and child to live in dignity and personal integrity, and to promote a set of basic principles that form the ethical and philosophical underpinnings of every society. These guiding principles are integrity, loyalty, accountability, transparency, independence, impartiality, tolerance, understanding, freedom from discrimination, gender equality, dignity and respect for different customs and cultures. Geoparks promote a

safe, inclusive, productive and supportive environment. Membership in the GGN means to abide a strict code of conduct. Individuals who experience inappropriate behaviour are encouraged to make a formal complaint to the appropriate bodies within the GGN.

Source: http://globalgeoparksnetwork.org/wp-content/uploads/2016/07/GLOBAL-GEOPARKS-NETWORK-CODE-OF-ETHICS-final.pdf (September, 2016).

Considering these principles, it is observed that there is great potential in geoparks to develop geoethics. Two examples of geoethics are presented, in which the scientists of a geopark can apply as an added value to the management activities of the geological resources that they carry out in the territory. According to Lucchesi and Giardino (2012), the principles of geoethics can be specified in the following actions for residents and visitors:

- To promote an attitude of "affection for the abiotic nature", not only through the transfer of scientific knowledge but also including the emotional dimension. For example, showing the beauty derived from the order and harmony of the crystalline structure of the minerals.
- Raise awareness about the dependency relationship between humanity and the planet. For example, as manifested in some cultures with the concept of "Pachamama-Mother Earth in Latin America". A consideration of our relationship with the Earth System that is less anthropocentric can facilitate a more balanced approach to the exploitation of natural resources.

2.9 Conclusions

The Earth is our only home: our well-being and our survival depend on its habitability and its resources (Matteucci et al. 2012).

The Anthropocene is a reminder that humans have influenced nature with a growing world population and our new technologies. In the twenty-first century, we face scarcity in critical resources, the degradation of ecosystem services and the erosion of the planet's capability to absorb our wastes. This situation is novel in its speed; however, we are the first generation widely aware of how our activities influence the Earth System, and also the first generation with the power to change our relationship with the planet.

Geoethics is expected to be a useful tool to achieve a new relationship with the geosphere, including the protection of its values involving society in geoconservation; a development strategy based on social, economic and environmental sustainability; an increase in the scientific geological culture; an improvement of the relationship with the scientific community, the media and those responsible for managing the territory.

As a model of geoethics commitment for professionals, a proposal (The Geoscientist's Promise) is proposed below. It is proposed by the Italian Commission of Geoethics (Matteucci et al. 2014) (box 3).

Box 3 The Geoscientist's Promise

- I promise I will practice geosciences being fully aware of the involved social implications and I will do my best for the protection of the geosphere for the benefit of mankind.
- I know my responsibilities towards society, future generations, and the environment for a sustainable development.
- In my job I will put the interest of society at large in the first place.
- I will never misuse my geological knowledge, not even under constraint.
- I will always be ready to provide my professional expertise in case of urgent need.
- I will continue to improve my geological knowledge lifelong and I will always maintain my intellectual honesty at work, being aware of the limits of my capabilities and possibilities.
- I will act to foster progress in geosciences, the dissemination of geological knowledge and the spreading of a geoethical approach to the management of land and geological resources.
- I will honour my promise that my work as a geoscientist or certified geologist will be fully respectful of Earth processes.

I promise.

Source: Formula proposed by the Italian Commission of Geoethics (Matteucci et al. 2014)

References

Almeida A, Vasconcelos C (2015) Geoethics: master' s students knowledge and perception of its importance. Res Sci Educ, 45. Springer, p. 889–906

Bohle M (2014) Recording the onset of the anthropocene. In: Lollino G, Arattano M, Giardino M, Oliveira R, Peppoloni S (ed) Engineering geology for society and territory, vol 7, Springer, Berlin, pp 161–163

Chalk R, Frankel MS, Chafer SB (1980) AAAS professional ethics project. American Association for the Advancement of Science, 1515 Massachusetts Ave. NW, Washington, D.C. 20005. AAAS Publication 80-R-4

Cordell D, Drangert J-O, White S (2009) The story of phosphorus: Global food security and food for thought. Glob Environ Change 19:292–305

Crutzen PJ (2002) Geology of mankind: the Anthropocene. Nature 415:23

Dowling RK (2011) Geotourism's global growth. Geoheritage 3:1–13

Eberhard R (1997) Pattern and process: towards a regional approach to national estate assessment of geodiversity. Australian Heritage Commission & Environment Forest Taskforce, Canberra, Australia
Eder W (2004) The global UNESCO network of geoparks. In: Proceedings of the First International Conference of Geoparks. Geological Publishing House, Beijing
Ellis EC, Haff PK (2009) Earth science in the Anthropocene: new epoch, new paradigm, new responsibilities. EOS Trans. 90(49):473
Ellis EC, Goldewijk KK, Siebert S, Lightman D, Ramankutty N (2010) Anthropogenic transformation of the biomes, 1700 to 2000. Glob Ecol Biogeogr 19(5):589–606
Foley SF, Gronenborn D, Andreae MO, Kadereit JW, Esper J, Scholz D, Pöschl U, Jacob DE, Schöne BR, Schreg R (2013) The Palaeoanthropocene – The beginnings of anthropogenic environmental change. Anthropocene 3:83–88
Gray M (2004) Geodiversity. Valuing and conserving abiotic nature. Wiley, Chichester
International Association for Geoethics—IAGETH (n.d.) http://www.icog.es/iageth/. Accessed 6 June 2017
International Association for Promoting Geoethics—IAPG (n.d.) http://www.geoethics.org/. Accessed 6 June 2017
Lovelock J (1979) Gaia. Oxford University Press, Oxford
Lucchesi S, Giardino M (2012) The role of geoscientists in human progress. Ann Geophys 55(3):355–359
MA (Millennium Ecosystem Assessment) (2005) Ecosystems and human well-being: Synthesis. Island Press, Washington, DC
Martínez-Frías J, González JL, Pérez FR (2011) Geoethics and deontology: from fundamentals to applications in planetary protection. Episodes 34(4):257–262
Matteucci R, Gosso G, Peppoloni S, Piacente S, Wasowski J (2014) The geoethical promise: a proposal. Episodes 37(3):190–191
Matteucci R, Guido G, Silva P, Sandra P, Janusz W (2012) A Hippocratic Oath for Geologists? Ann Geophys 55(3)
McKeever P (2013) Memory of the Earth. World Sci 11(3):4–13
Monastersky R (2015) The human age. Nature 519(7542):144–147
Newsome D, Dowling RK (2010) Geotourism: the tourism of geology and landscape. Goodfellow Publishers, Oxford
Nikitina N (2012) Geodiversity, and the geoethical principles for its preservation. Ann Geophys 55(3)
Nikitina N (2016) Geoethics: theory, principles, problems. In: Monograph, 2nd edn. Geoinformmark, Ltd., Moscow. 256 p
Palsson G, Szerszynski B, Sörlin S, Marks J, Avril B, Crumley C, Hackmann H, Holm P, Ingram J, Kirman A (2012) Reconceptualising the "Anthropos" in the Anthropocene: integrating the social sciences and humanities in global environmental change research. Environ Sci Policy 1–11
Pascual JA (2011) La coevolución de la Tierra y de la vida o cómo se han inflido mutuamente la geología y la vida. Alambique 67:37–45
Peppoloni S, Di Capua G (2012) Geoethics and geological culture: awareness, responsibility and challenges. Ann Geophys 55:335–341
Peppoloni S, Di Capua G (2015) The meaning of geoethics. In: Peppoloni S, Wyss M (eds) Geoethics: ethical challenges and case studies in earth science. Elsevier, Waltham, Massachusetts
Prosser C (2002) Term of endearment. Earth heritage 17:12–13
Ruddiman WF (2003) The anthropogenic greenhouse gas era began thousands of years ago. Clim Change 61:261–293
Savolainen K (1992) Education and human rights: new priorities. In: Adult education for international understanding, human rights and peace. Report of the Workshop held at UIE, Hamburg, 18–19 April 1991, UIE Reports, n. 11, p 43–48

Sharples C (1995) Geoconservation in forest management: principles and procedures. Tasforest 7:37–50

Smith BD, Zeder MA (2013) The onset of the Anthropocene. Anthropocene 4:8–13

Steffen WA, Persson L, Deutsch M, Williams J, Zalasiewicz C, Folke J, Rockström C, Crumley P, Crutzen L, Gordon M, Molina V, Ramanathan K, Richardson M, Scheffer J, Schellnhuber J, Svedin U (2011) The Anthropocene: from global change to planetary stewardship, Working Paper No. 2. Prepared for the "3rd Nobel Laureate Symposium on Global Sustainability: Transforming the World in an Era of Global Change", in Stockholm, 16–19 May 2011. Stockholm Resilience Centre, the Royal Swedish Academy of Sciences, the Stockholm Environment Institute, the Beijer Institute of Ecological Economics and the Potsdam Institute for Climate Impact Research

von Bertalanffy L (1968) General system theory: foundations, development, Applications. George Braziller, New York

Zouros N (2004) The European Geoparks Network – geological heritage protection and local development. Episodes 27(3):165–171

Part II
Argentina

Chapter 3
Towards a Latin American Geoethics?

Cristian Lorenzo

Abstract Ecological disasters are taking place in Latin America as a consequence of the development of extractive industries and the rising demand of commodities by international markets. The role of Latin America in the world system provides an insight into the concept of Geoethics. This article argues on the existence of theoretical and historical backgrounds in Latin American ideas to begin a discussion for the elaboration of a "Latin American Geoethics".

Keywords Latin American Geoethics · Philosophy · Natural resources Development · International relations

3.1 Introduction

The objective of this article is to "unthink" the concept of Geoethics. Nowadays, ecological disasters in Latin America take place as result of the increasing exploitation of natural resources and subsequent exportation to different markets. In last November, an environmental disaster occurred in Brazil when two dams, containing wastewater from iron industries, collapsed in the state of Minas Gerais. Their rupture caused a toxic mudslide that polluted the Rio Doce and caused deaths in Mariana city. Two months before, an environmental tragedy had happened in Argentina. Barrick Gold, a Canadian mining company, admitted it had problems with its pipelines. In fact, cyanide solution had spilled into the nearby river and caused environmental as well as health problems to people living in the area. As a consequence, the provincial government of San Juan suggested to its people to avoid or reduce the water consumption due to the pollution of its rivers. These two cases have something in common: they reveal the environmental cost of extractive

C. Lorenzo (✉)
Centro Austral de Investigaciones Científicas—Consejo Nacional de Investigaciones Científicas, Instituto de Ciencias Polares, Ambiente y Recursos Naturales—Universidad Nacional de Tierra del Fuego, Bernardo Houssay 200, 9410 Ushuaia, Argentina
e-mail: clorenzo@conicet.gov.ar

R. D. Acevedo and J. M. Frías (eds.), *Geoethics In Latin America*,
The Latin American Studies Book Series,
https://doi.org/10.1007/978-3-319-75373-7_3

industries in Latin America as well as the responsibilities of professionals involved in those situations.

This article asserts that there are theoretical and historical backgrounds in Latin American ideas to begin a discussion on a "Latin American Geoethics". It will be divided as follows. First, an overview of places in which Geoethics knowledge circulates will be presented. Then, Latin American ideas will be analysed to assess the possibility of establishing a dialogue on Geoethics between Social Sciences and Humanities on one hand, and Natural Sciences, on the other. Finally, conclusions and new questions raised will be presented for further research.

3.2 Unthinking Geoethics

Two years after the fall of the Berlin Wall, Wallerstein (1991) published the book "Unthinking the Social Sciences". One of its most important contributions is the distinction between two concepts: "unthink" and "rethink" the situation of Social Sciences (Wallerstein 2001: 3). In spite of seeming to be a pun, they have different meanings and implications. It may be noted that to rethink is common to scientific and academic fields, in particular, when new evidence appears questioning the prevailing theories. However, the idea of "unthinking" highlights the existence of "misleading" and "constraining" assumptions that shape a mentality.

A possible exercise for unthinking is the identification of those places where knowledge—related to Geoethics—circulates. The essential character of this concept refers to an ethical issue involved in the professional activities of geologists. Geoethics was born in the 1990s; Nemec Václav, who was then the Vice President for the International Association for Promoting Geoethics (AGID), was considered as the founder of this discipline. Since then, Václav was relevant for the institutionalization process that took place. Different actions needed to be considered at this point. Geoethics was an important issue in international congresses; working groups were formed to address different aspects of the Geoethics, with the support of the institution recently mentioned. Moreover, it is important to highlight that Geoethics was included as an area of the Geosciences Journal and the International Statement on Geoethics was approved in 2011 within the framework of a meeting at AGID (González and Martínez-Frías 2011).

Regarding this process, the literature was analysed to find places in which Geoethics knowledge circulated. Two places were identified in this exercise: the professional and the social orbits.

3.2.1 *Professional field*

One of the axes of the discussion about Geoethics refers to the institutionalization of ethics principles for geology-related activities. This ethical code implies a set of themes: land use and management; the role of science in society, the earth protection; and finally, fostering public environmental awareness and an attitude towards a lifelong learning (Matteucci et al. 2012).

These principles governing the ethical code are not defined in the abstract; it is necessary to have a wider context for its interpretation. In this sense, the economic crisis affecting the European Union is one of the key issues to take into account. An editorial in the Journal of Geological Information "Land and Technology" reveals the critical employment situation for geologists in Spain. For this reason, they say that "are becoming a legion of geologist who makes bags and emigrate" (Ilustre Colegio Oficial de Geólogos 2011a: 2). The other dimension involved in the development of mining industries in Latin America as an area with potential for professional development of European geologists (Ilustre Colegio Oficial de Geólogos 2011b: 61).

3.2.2 *Social Sciences and Humanities fields*

In the context of the development of Geoethics, its relation with Social Sciences and Humanities is crucial. In line with this consideration, an interview to Franco Ferrarotti by Peppoloni can be specially considered (Peppoloni 2012). This sociologist argues that geologists have no visibility as an actor in the Italian public domain. In spite of this, Ferrarotti considers that they have an important role in society. He specifically highlights that the intervention of geologist in public opinion could be a relevant contribution to society. Ferrarotti has a critical view on mass media because he considers they are ethically indifferent and permanently dramatize what happens in reality.

3.2.3 *Social field*

The development of the Geoethics also suggests the recognition of the place that the academic scientific community has in the society. It involves a set of matters. For example, the definition of the scope of geologist professional activities. Peppoloni and Di Capua define Geoethics as a discipline for "the study and promotion of the evaluation and protection of the geosphere" (Peppoloni and Di Capua 2012). In the delimitation of its responsibilities, they define a set of issues that go beyond the merely technical aspects. In that sense, the connections of geosciences and the press are a key factor to consider within this vision on what implies the activities of a geologist.

According to Limayela, Geoethics has a defined social role. In his view, Geoethics is an incipient discipline with potentiality to govern the social interactions in the community; in other words, this vision considers that it could be a cooperation framework for different social stakeholders: state, industries, civil society and geoscientists. He puts forward three relevant domains for Geoethics:

education, dissemination of scientific knowledge and conservation of geological heritage (Limaye 2015).

The visibility of geologist activities in mass media is another relevant aspect involved, in terms of the vision held by the society on geologists. As an example, geologist in Spain outsourced its strategy of communication through a consultant to gain more visibility in Spanish newspapers. In accordance with the "Ilustre Colegio de Geólogos" from that country, 2011 had a positive balance. Its mentions in newspapers, radio and television represented an increase of 137% in comparison with the previous year. Earthquakes, volcanoes and nuclear power plants were the most recurrent themes. It should be noted that this year was special because there was an earthquake in Japan, an earthquake in Lorca, and there was also a volcanic eruption in the island of Hierro in Spain (Ilustre Colegio Oficial de Geólogos 2011c).

To sum up, Geoethics display two fields in which its knowledge circulates: professional and the social orbits. However, the scope of Geoethics in others fields is not considered here as a finished process. It should be remembered a quote from German philosopher Arendt that is significant at this point: "the task of thinking is like the work of Penelope, that every morning undid what he had done the night before" (Arendt 2008: 117). Taking up this idea, the purpose of the next section is to promote a dialogue between Geoethics and Social Sciences and Humanities in Latin America, boosting a new redefinition of Geoethics.

3.3 Reasons for a Latin American Geoethics?

This section addresses relations between Geoethics and Social Sciences and Humanities in Latin America. Within this framework, the key question is the following: are there enough reasons for suggesting the existence of a Latin American Geoethics? There are mainly two reasons to consider: (i) it will represent an expansion of Geoethics, opening a dialogue with Latin American ideas on the environment and natural resources; and (ii) Latin America is an important region for the professional development of geologist worldwide (Ilustre Colegio Oficial de Geólogos 2011b: 61–63). These arguments provide different insights to Latin America: as a theme, as a place and as a source of thinking (Seitz 2012: 30). To address this question, ideas elaborated in Latin America from Social Sciences and Humanities will be addressed as follow.

3.3.1 Latin American Ideas

Before proceeding, it is necessary to have a clear definition of what Latin American philosophy implies. Fernández Nadal states that first of all, it is necessary to have a recognition of a Latin American tradition of thinking. In addition, this author highlights certain common characteristics. Thinkers usually refer to two dimensions closely linked and present in their way of reasoning: the historical context matters as well as the empirical situation involved in a community (Fernández Nadal 2008: 232–237). It is important to highlight another overall feature: the Latin American philosophy is elaborated outside the centre of power, like Africa or Asia (Devés Valdés 2008: 400–401).

There are several authors within Latin American philosophy to be considered. According to this diversity, this section will select some of them, in particular those whose contribution enrich the discussion on the possibility of a Latin American Geoethics. So, the aim is not to do an exhaustive literature review; in contrast, this section will prioritize categories from different authors that show diverging insights to the aforementioned question.

The first author is Arturo Roig. According to this author, the identification and recognition of the place of enunciation are not irrelevant. This statement has a political dimension: behind any discourse, there is always a subjectivity. This idea is mainly important to track the genealogy of any discourse. Roig puts forward the category of "anthropological a priori", retaking an idea originally posed by Hegel in the Introduction to the history of philosophy. This Argentinean author argues that the beginning of philosophy occurs when a philosopher affirms itself. This is the starting point of a Latin American vision to think about what happens in the world. Furthermore, the philosophy proposed by Roig is not for doing after the events; on the contrary, it believes that philosophy is a kind of knowledge that becomes critical before events happen (Roig 2009: 15).

Alejandro Auat is another author within the Latin American tradition of thinking. He published a book called "Towards a located political philosophy", in which he addressed the character situated of any thinking. For this author, the "located philosophy" implies to consider both a conceptual level and the historical grounds, studying its permanent feedbacks. This author argues that it is possible to go beyond the "universal-located" tension, and from this perspective, it will be possible to understand the meaning of its parts defining larger entities (Auat 2011: 80).

Kusch distinguishes the action of thinking from the philosophy. The recognition of this difference has consequences on the knowledge elaborated. For thinking, it is not necessary to have an encyclopaedic accumulation of data; in spite of this, it implies the attempt of having a comprehension of totality in an unsystematically way. The act of philosophize refers to the intellectual activities in an academic field, and this implies the existence of a set of rules that gives predictability and defines standards for the quality of the knowledge elaborated. Instead of referring to a universal thinking, Kush argues the character deterministic of the place in any discourse (Kusch 2008).

To sum up, Roig, Auat and Kusch are authors that offer insights to rethink Geoethics from Latin America. They share the idea of having a critical viewpoint on universal discourses, taking into account the importance of the place of enunciation. For such reason, Geoethics could be considered as a predicate and the geoscientist in the international community as a form of subjectivity. So, the next question is: are there additional elements in the Latin American literature on natural resources to assess the idea of a Latin American Geoethics?

3.3.2 Structural Historical Features

In accordance with the Latin American philosophical attitude, recently mentioned above, the purpose of this section is to characterize the structural historical characteristics of Latin America through the literature available. To have a big picture of its situation, it is necessary to consider the abundance and diversity of its natural resources (Unión de Naciones Suramericanas 2013) as well as the existence of different problems that shape an environmental crisis, such as pollution, depletion of natural resources, demographic explosion and destruction of the ozone layer (Estenssoro Saavedra 2011: 10).

Several authors pointed to the current model of Latin American development: highlighting the essential logic and its limits. Just for mentioning some of them, Gudynas identifies actors involved in the phenomenon that he calls "extractivism". This concept implies the mining and oil as well as the agriculture. In all of those sectors, the logic is the same: there is an intensive exploitation of natural resources for exports. This is the current role of Latin America in the world. In addition, Gudynas highlights that state and transnational corporations are involved in this model of development. In addition, environmental conflicts were part of its processes, showing the limits of this way of understanding development (Gudynas 2015).

Svampa, meanwhile, posits that there was a process in which Latin America is governed for "Commodities consensus" (Svampa 2013). This category implies to have a special consideration for the exports of primary goods on a large scale. This idea also refers to a new international order in which the economic and political conjuncture is characterized by the "boom" of raw materials' international prices. As for the mining sector, she has a critical point of view: "there is no Latin American country in which mining project by large scale has no social conflicts between mining companies and the government against communities" (Svampa 2013).

Estenssoro Saavedra provides an additional insight. He considers that the global environmental problem is essentially a political matter, apart from having different impacts on nature. He highlights that the human being is not precisely the cause of it, but a particular group of them is responsible for this. For this reason, it is possible to state that the environmental crisis is socially rooted and it is a consequence of the existence of power relations at international level. As approaches for tackling this global problem, he recognized the existence of different perspectives and subsequently, conflict of interests (Estenssoro Saavedra 2014).

Saguier has an additional point of view for taking into account different characteristics of Latin America that are relevant for the comprehension of environmental socio-conflicts. He argues that multinational companies such as mining and oil violate human rights. At an international level, the Permanent People´s Tribunal has the political role of addressing cases in which human rights are violated and also invisible for the rest of the society. Since 2006, this forum has been used by different stakeholders in Latin America (environmental groups, indigenous movements, peasants and workers) to denounce transnational companies violating human rights. Apart from those stakeholders that put multinational companies in the dock,

Saguier argues the necessity of having a wider perspective due to the existence of a model of complicity between transnational corporations, states and international financial institutions (Saguier 2010).

3.4 Conclusions

As it was stated before, the main objective of this article was to unthink Geoethics. In line with these, this article shows that there are theoretical and historical backgrounds in the ideas of Latin America to open a discussion for the elaboration of a "Latin American Geoethics".

The significance of Geoethics is still under construction. As a part of an open process, different fields were identified in which knowledge on Geoethics circulates. In accordance with this line of interpretation, it is possible to sketch a map of its scope: a level for the professional activities, a level for Social Sciences and Humanities; and finally, a level for the society itself. Within this framework, the aim of the key question was going beyond the current relations between Geoethics, Social Sciences and Humanities. For such reason, a set of categories elaborated from Latin American authors were considered on structural characters of natural resources in Latin America.

Roig, Auat and Kusch posit a critical vision to universal discourses as well as they contributed with different concepts to open a discussion on the possibility of a Latin American Geoethics. Later, different perspectives on the natural resources situation in Latin America were taken into account. Gudynas, Svampa, Estenssoro Saavedra and Saguier showed the role of Latin America in the world, conditioned by power relations. Those authors give different elements to have different insights into the discussion on the meaning of Geoethics from the Latin American perspective.

Finally, it is expected that results of unthinking Geoethics were the starting point to deepen the enunciation spaces and enrich the international process in which its future is discussed. Moreover, two new questions arise: is the "buen vivir" paradigm from Latin America a way of enriching the international discussions on Geoethics? And finally, what is the role of Geoethics in the current world historical system?

References

Arendt H (2008) De la historia a la acción. Editorial Paidós, Buenos Aires, 171 pp

Auat A (2011) Hacia una política situada. Waldhuter Editores, Buenos Aires, 236 pp

Devés Valdés E (2008) "Pensamiento periférico", en Biagini, Hugo; Roig, Arturo. Diccionario de pensamiento alternativo. Editorial Biblos, Buenos Aires, 400–401

Estenssoro Saavedra F (2011) Introducción; en Cubillos, Adela; Estenssoro Saavedra, Fernando (comp.) Energía y Medio Ambiente. Una ecuación difícil para América Latina. Los desafíos del crecimiento y desarrollo en el contexto del cambio climático. Instituto de Estudios Avanzados de la Universidad Santiago de Chile. Santiago de Chile, pp 9–21

Estenssoro Saavedra F (2014) Historia del debate ambiental en la política mundial 1945–1992. La perspectiva latinoamericana. Instituto de Estudios Avanzados de la Universidad de Santiago de Chile. Santiago de Chile, 215 pp

Fernández Nadal E (2008) "Filosofía latinoamericana", en Biagini, Hugo; Roig, Arturo. Diccionario de pensamiento alternativo. Editorial Biblos, Buenos Aires, 232–237

González JL, Martínez-Frías J (2011) Geoethics: un reto para la deodontología profesional, Tierra y tecnología, 40: 10–14

Gudynas E (2015) Extractivismos. Ecología, economía y política de un modo de entender el desarrollo y la naturaleza. Ediciones CEDIB. Cochabamba (Bolivia). 453 pp

Ilustre Colegio Oficial de Geólogos (2011a) "Condenados a emigrar". Revista de Información Geológica Tierra y Tecnología, 41: 2

Ilustre Colegio Oficial de Geólogos (2011b) "Chile es el paraíso para los profesionales en la minería", Revista de Información Geológica Tierra y Tecnología, 41:61–63

Ilustre Colegio Oficial de Geólogos (2011c) El Colegio de Geólogos consigue 1.973 impactos en prensa durante 2011, un 173% más que el año anterior. Revista de Información Geológica Tierra y Tecnología, 40:64–66

Kusch R (2008) La negación en el pensamiento popular. Editorial Las Cuarenta, Buenos Aires. 220 pp

Limaye, SD (2015) A geoethical approach to industrial development of georesources and groundwater use: the Indian experience. Geological Society London Special Publications, 419

Matteucci R, Gosso G, Peppoloni S, Piacente S, Wasowski J (2012) A Hippocratic Oath for geologists? Ann Geophys 55(3):365–369

Peppoloni S (2012) Social aspects of the Earth sciences - Interview with Prof. Franco Ferrarotti. Ann Geophys 55(3):347–348

Peppoloni S, Di Capua G (2012) Geoethics and geological culture: awareness, responsibility and challenges. Ann Geophys 55(3):335–341

Roig A (2009) Teoría y crítica del pensamiento latinoamericano. Editorial Una Ventana, Buenos Aires, 380 pp

Saguier M (2010) En el banquillo de los acusados: Empresas transnacionales y violaciones de derechos humanos en América Latina. Revista de Negocios Internacionales 3(3):116–153

Seitz A (2012) Conocimiento e integración: remozando las relaciones internacionales desde América Latina, sus recursos naturales y sus dilemas democráticos. In: Theiler, Julio César; Agramunt, Luis Felipe. La enseñanza europea para la integración latinoamericana. Presentaciones del II Congreso Internacional de la Red de Integración Latinoamericana. Ediciones UNL, Santa Fé, pp 29–42

Svampa M (2013) Consenso de los Commodities y lenguajes de valoración en América Latina. Revista Nueva Sociedad 244:30–46

Unión de Naciones Suramericanas (2013) Potencial de recursos energéticos y minerales en América del Sur. Coincidencias jurídicas hacia una Estrategia Regional. OLADE, Quito, 140 pp

Wallerstein I (1991) Impensar las Ciencias Sociales. Límites de los paradigmas decimonónicos. Editorial Siglo XXI, México, 114 pp

Chapter 4
The Superior Geology Professional Council and Ethics

Claudio Alberto Parica

Abstract Argentina since 1963 started early to manage ethics and geoethics in the country through the Superior Geology Professional Council. The newest Code of Ethics was performed in 2003, in agreement with new laws in the country about ethics. The Council has the challenge to observe the fulfilment of ethical rules of the Earth Science for nationals and foreign professionals who develop tasks in the country in benefit of the community.

Keywords Argentine geologic professional council · History · Code of Ethics

4.1 Introduction

Since 1963 because the Law 19937, the Superior Geology Professional Council exists in the Argentine Republic. Pioneer colleagues decided to buy a building in Buenos Aires downtown called "The Geologist House," where the administration of the Council operates. The institution promotes the exercise of the profession within a framework of dignity and ethics, based on the law of professional practice.

After more than 50 years, a lot of goals were obtained, and one of the most important is that Geology was declared a Career of Public Interest, with the addition to be regulated into normative of excellence to get the final title. All the argentine universities with Geology must be in agreement with Minister Regulations (Resolution 1412/2008, in this way, every university receives grants for equipment.

The organization of Geology exercise started just in 1963, with the condition for any geologist in the country to get the licence in our national Council. Later, because the federal condition of Argentine, some provinces created local Councils (Córdoba, San Juan, Mendoza, Jujuy, Catamarca, La Rioja, San Luis, Chubut, La Pampa, Buenos Aires Province, Tucumán and Salta). Between all the councils in

C. A. Parica (✉)
Consejo Superior Profesional de Geología, CABA,
Maipú 645, C1006ACG, Aires, Argentina
e-mail: cparica@gmail.com

R. D. Acevedo and J. M. Frías (eds.), *Geoethics In Latin America*,
The Latin American Studies Book Series,
https://doi.org/10.1007/978-3-319-75373-7_4

the country, an agreement about licences was signed, where every licence is accepted at any other jurisdiction.

In this way, the exercise of Geology is clearly organized all around Argentina. The respect to persons, human rights, and honest job is a priority for the CSPG.

4.2 Some Facts About Ethics

The Council at several opportunities might to take some problems to solve because different behaviours of professionals coming from other areas of science or pseudo sciences. Frequently, in these kinds of situations, the Council was present to take the defence of the Geology, geologists and its incumbencies. The knowledge of a science and its attributes is the right tool to protect the environment, the inhabitants of the country.

There is a particular case in which the Council had to decide about the behaviour of a college against some other colleges. This case was very sensitive, because the college alive starts an attack against death colleges, without any possibility of any defence. This case started with a note into a Magazine of the Buenos Aires University; the alive college exposed a lot of disagreeable concepts against colleagues died several years ago.

4.3 Consequences

This alive colleague got a punishment from the Council because of his behaviour, because of the treatment without any respect for death colleague without any possibility of defence. The punishment was announced at first in the Official Bulletin for the Argentine Republic, also for Buenos Aires City Official Bulletin. This punishment did not imply money, just to put his name on evidence to the community.

Subsequently the ethically sanctioned colleague has had legal consequences, losing a trial in the justice of the Argentine Republic. This example is presented to the students of Geology as a fact worthy of repudiation.

Here is presented the Geology Code of Ethics for the Argentine Republic.

4.4 The Code of Ethics

Chapter I
On competence and observance

Section 1 The basic concepts and standards of this Professional Code of Ethics shall be obligatorily observed by all geologists who practice their profession in the territory of the Argentine Republic, which includes the observance of those rules of behaviour which—although not set forth in the present code—are mandatory since they protect the honour and dignity of the profession.

Chapter II
On the general principles of professional ethics

Section 2 Within the scope of application of this code, professional ethics is a set of the best moral concepts and criteria established to guide the conduct of professional geologists in the practice of their knowledge, in such a way that their actions can be of benefit to the society.

Section 3 The professional conduct of geologists shall be governed by:

a. The proper application of the rules of Art and Science for which they are authorized.
b. Integrity, impartiality, loyalty and confidentiality.
c. Respect for the work of others.
d. The purpose of enhancing the prestige of the profession.

Chapter III
On the obligations imposed by professional ethics.

Section 4 In their professional practice, geologists shall be required:

a. To observe and ensure the fulfilment of the legal standards that govern the practice of the profession and to report to the competent authorities in case of becoming aware of an act of infringement.
b. To refrain from advising or participating when that implies favouring or facilitating illegal acts.
c. To refrain from participating in activities or appearing in some documents together with individuals or institutions that unduly practice the profession.
d. To refrain from providing or offering services which exceed their professional competence or evade the rules of Art and Science.
e. To verify the truthfulness of the information supporting their opinions and evidence and to express them in a precise and complete manner, in such a way that it may not lead to erroneous or ambiguous interpretations.
f. Not to sign or certify those documents or reports which have not been personally prepared or supervised.
g. To properly quote the opinion of third parties and to avoid plagiarism.
h. Not to consent the illegal and/or unethical practice of the profession under his/her hierarchical or working responsibility.
i. To report the Superior Council on Professional Geology if he/she became aware of any act of infringement of the standards set forth in this Code of Ethics.

Chapter IV
On violations of professional ethical standards

Section 5 The geologist who fails to observe one or more of the principles and obligations set forth in this code shall commit an ethical violation.

Section 6 The Court of Professional Ethics shall classify ethical violations in terms of whether they are minor or major offences; they shall bear the supplementary penalty of the disqualification integrate the Court of Ethics. The disqualification would be between two and four years in the case of minor offences and permanent for major ones.

Section 7 All complaints on ethical issues shall be received up to 5 years after the offence was committed.

Section 8 For the purpose of qualifying the nature of the penalty, the repeated offences committed by a professional geologist shall be considered as aggravating facts.

Chapter V
About the rules of procedure

Section 9 Once a complaint against a professional geologist is placed, the president of the Superior Council on Professional Geology shall call the Court of Professional Ethics and shall submit all the records received. The Court shall decide whether the case at issue is related to professional ethics and, if it were, it shall arrange to continue with the proceedings.

Section 10 Should the Court of Ethics consider that the case at issue is not related to professional ethics, it shall immediately report about it to the president of the Council so that he/she can communicate it to the informant and the defendant. The informant shall be able to appeal the resolution before the Superior Council on Professional Geology up to 30 consecutive days starting on the day of the notification.

Section 11 Once the resolution that determines the continuation of the proceedings has been consented and/or becomes final, the Superior Council on Professional Geology shall forward the complaint to the defendant, who shall have to give an answer within the term of 15 consecutive days in case he lived in the Ciudad Autonomic de Buenos Aires or 30 consecutive days if his/her residence were in the interior of the country. The complaint shall be forwarded by registered letter, including a copy of the written accusation.

Section 12 The Court of Ethics shall study the evidence that both parties could have submitted in a period of 30 consecutive days, but if there were some additional evidence, it shall be possible to extend the term up to 60 consecutive days.

Section 13 The Court of Ethics shall submit the decision and the final conclusions to the consideration of the Superior Professional Council on Geology and shall also deliver the proceedings.

Section 14 Once the Superior Council on Professional Geology has received the records and the Court's decision, it shall issue the final judgment and it shall make the relevant communications.

Section 15 Within a period of ten days following the notification, it shall be possible to appeal the judgment. In the case of a major offence, it shall be possible to lodge a revocation appeal before the incumbent National Judge in Federal Administrative Matters within a period of ten days following the notification of the judgment disregarding the revocation.

4.5 Conclusions

Argentine is one of the pioneer countries which considered the ethics for Geology, from the very beginning with creation of the Superior Geology Professional Council, the observance of ethics was a priority. There were a few cases during the 53 years of existence of faults against ethics coming from geologists, some others coming from people without the right incumbencies, where the Council took legal actions. We expect a best future with a strong compromise on the ethics in Geology.

Part III
Brazil

Chapter 5
Ferruginous Geosystems and the Current Situation of Iron in Brazil: Economic Growth or Environmental Regression?

Giana Márcia dos Santos Pinheiro

Abstract In a recessive economic scenario, which imposes or exacerbates strong political and institutional pressures to Brazil's environmental agenda, the geology practice in the country is, both in the academic and in the scientific fields—and above all, in its practical and professional applications, especially concerning the iron ore production—, ethically challenged to give answers that may dictate the fate of distinct spaces and crucial ecosystems. In this context, it is here illustrated that decisions that currently fall on the geological exercise may affect the present and future existence of an extraordinary diversity of life-forms, both human and non-human. The presentation that follows highlights the clear ethical dimension of the challenge that lies over the professionals of the country, for it reveals the inseparability between the technical and the political dimensions of the geological craft and points out that the geology science fulfills a central role for the environmental conservation (or degradation) of *Ferruginous Geosystems*.

Keywords Ferruginous Geosystems · Iron mining · Protected areas
Water security · Environmental policies in Brazil

5.1 Introduction

This contribution seeks to characterize the distinct aspects of importance, regarding the environmental conservation, of the so-called *Ferruginous Geosystems*. These complex formations, which face powerful interests linked to mining industries—especially the iron one—have been subject to assaults that endanger several biodiversity hotspots (as well as the geological, biological, and cultural heritages), which perform crucial environmental services that may be compromised. Such geosystems

---

G. M. dos Santos Pinheiro (✉)
University of Brasília (UnB), Brasília, Brazil
e-mail: giana@unb.br

G. M. dos Santos Pinheiro
University of Toulouse III, Paul Sabatier (UPS), Toulouse, France

R. D. Acevedo and J. M. Frías (eds.), *Geoethics In Latin America*,
The Latin American Studies Book Series,
https://doi.org/10.1007/978-3-319-75373-7_5

are exposed, in Brazil, to an environmental degradation that has many adverse consequences, particularly related to water availability and the guarantee of water multiple uses, critical aspects regarding the climate change projections that are already felt on a global scale, and should be felt even more in the coming decades. To support the subsequent exposure, several studies listed in the most important book published about this particular type of formation, entitled *Ferruginous Geosystems in Brazil* (*Geossistemas Ferruginosos do Brasil* in Portuguese—Carmo and Kamino 2015a)[1] were accessed. After a presentation of some general considerations about iron and *Ferruginous Geosystems*, this paper will provide an overview of the threats that, due to significant economic and governmental interests, weigh on these systems. Once the trans-specific rights to a balanced environment—for both the current and the future generations—are at stake, the ethical dimension associated with the exercise of geology can be emphasized.

5.2 Iron

Iron is the most abundant element (by mass) on Earth, forming much of its inner core, as well as the inner core of many other planets, accounting for approximately 35% of their total masses. It is the second most abundant metal on Earth (behind aluminum) and fourth most abundant element in the Earth's crust (~5.1%). It is found in almost all rocks, soils, and hundreds of minerals (Nielsen 1960), and it also forms the nucleus of red giant stars, representing one of the most abundant elements in the Universe.

The most common iron redox states at the Earth's surface are ferrous iron (Fe^{2+}) and ferric iron (Fe^{3+}), but this element can also exhibit the oxidation states +4 and +6 in unstable salts (Nielsen 1960). In the presence of free oxygen, iron is under the oxidized form Fe^{3+}, which has very low solubility at neutral pH (Curie and Briat 2003) and is mostly present in the form of iron (hydr)oxides, such as ferrihydrite ($FeOOH \cdot 0.4H_2O$, Zhao et al. 1994), goethite (α-FeOOH), hematite (α-Fe_2O_3), maghemite (γ-Fe_2O_3), and magnetite (Fe_3O_4, formed by both Fe^{2+} and Fe^{3+}, Cornell and Schwertmann 2003). Other common natural minerals that contain iron are carbonates (e.g., siderite—$FeCO_3$—and ankerite—$Ca(Mg,Fe)(CO_3)_2$), silicates (e.g., greenalite—$(Fe^{+2},Fe^{+3})_{2\text{-}3}Si_2O_5(OH)_4$), and sulfides (e.g., pyrite—FeS_2).

According to Ruchkys (2015), the main iron deposits that occur on the planet are associated with iron formations (IFs), which are metasedimentary rocks with more than 15% of iron in their chemical compositions (Souza and Carmo 2015). The IFs include rock types such as jaspilites and itabirites (both banded iron formations—BIFs—, respectively of sedimentary chemical and metamorphic origin), and other

[1]Available for full download at the Pristino Institute Web page—http://www.institutopristino.org.br/wp-content/uploads/2016/03/Geossistemas-ferruginosos-no-Brasil-CD.pdf (accessed February 2018)

iron-rich rocks (Ruchkys 2015). In this situation, the ferruginous laterites, also called ferruginous crusts, lateritic crusts, or *cangas* (Sect. 5.4), which are primarily formed by chemical weathering, are noteworthy (Ruchkys 2015).

The term *Ferruginous Geosystems* (FGSs) is here used to describe all the elements that constitute spatial units whose lithological substrate consists of ferruginous rock types, including its physical components (i.e., mineralogical, hydrological, soils, and all the relationships between them) and in particular the *cangas* that generally overlie such rock types (Souza and Carmo 2015; Ruchkys 2015).

The FGSs show, as notable features, the antiquity and the heterogeneity of the constituent rocks (which are often BIFs from the Archean—older than 2.5 billion years—and from the Paleoproterozoic—2.5 to 1.8 billion years) and the presence of associated *cangas*. The enormous diversity and the complex evolution of these *cangas* are factors that favor the biodiversity development and its maintenance (Trendall and Morris 1983; Monteiro et al. 2014; Salgado and Carmo 2015; Carmo and Kamino 2015b; Souza and Carmo 2015). Among the different ecosystems found in Brazilian biomes, the FGSs are the least known and the most threatened (Ferreira et al. 2015). This status is due to their restricted distribution and common association with the main iron ore deposits of the country (Jacobi and Carmo 2008; Ferreira et al. 2015).

5.3 Banded Iron Formations (BIFs) and Conservation Units in Brazil

Brazil concentrates one of the largest occurrences of IFs (Rosière and Chemale Jr. 2000; Bekker et al. 2010). These IFs include BIFs, diamictites, dolomites, and ferruginous phyllites, as well as other rocks with predominant iron composition (Souza and Carmo 2015). The BIFs occur, in many cases, as topographical features that behave as terrestrial islands, making them ideal systems to investigate the importance of connectivity (or isolation) and evolutionary processes (Gibson et al. 2015).

Although the mining industry shows a small footprint compared to industries such as pastoralism, the flora and fauna complexity in BIF areas can harbor unusual and diverse biotas (e.g., Jacobi et al. 2007; Gibson et al. 2010, 2012, 2015; Pepper et al. 2013 and references in Carmo and Kamino 2015b). Ecological interactions are still not well studied in FGSs (Jacobi and Carmo 2008, 2011; Jacobi et al. 2015) and researches to date focused on a small number of BIF endemics currently or potentially being impacted by mining operations (Gibson et al. 2015).

Brazil, China, and Australia are the world's major iron ore producers (Gibson et al. 2015), and although the demand for this element has decayed in the past years, it is still high, especially from China. The main Brazilian iron reserves are located in only three states (Souza and Carmo 2015), called Minas Gerais, Mato Grosso do Sul, and Pará. Six Brazilian FGSs stand out as important iron ore deposits

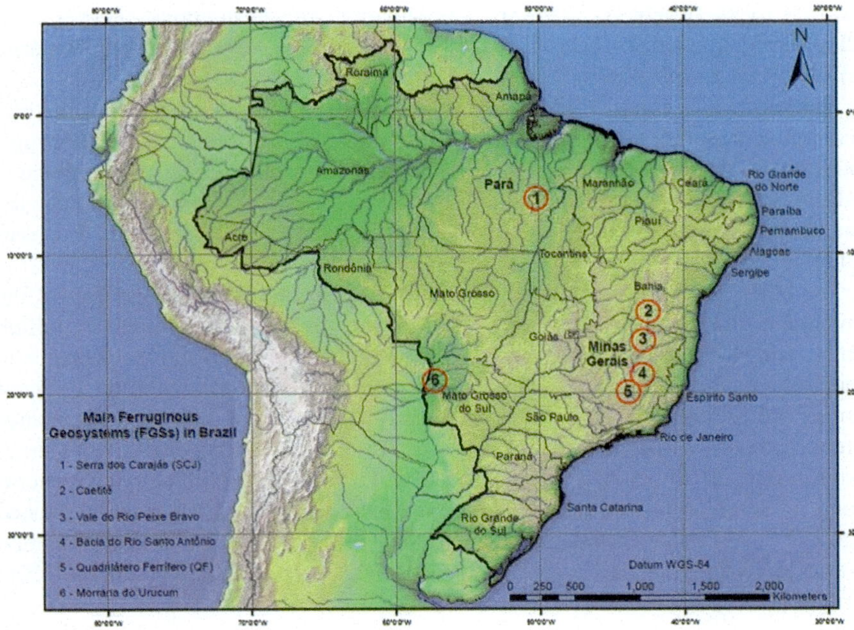

Fig. 5.1 Location of the six main *Ferruginous Geosystems* (FGSs) in the country: (1) Carajás Range (SCJ)—state of Pará; (2) Caetité—state of Bahia; (3) Peixe Bravo River Valley—state of Minas Gerais; (4) Santo Antônio River Basin—state of Minas Gerais; (5) Ferriferous Quadrangle (QF)—state of Minas Gerais; and (6) Morraria do Urucum—state of Mato Grosso do Sul. The circles represent the regions where the FGSs are inserted. For specific limits of these systems, please refer to Carmo and Kamino (2015a, b). Political map of Brazil showing all the 26 states (and the Federal District—DF) and the main Brazilian rivers. Digital Elevation Model (500 meters, SRTM), LANDSAT mosaic. *Credit* G. M. dos Santos Pinheiro

(Fig. 5.1): in the state of Minas Gerais, the Ferriferous Quadrangle (or *Quadrilátero Ferrífero*—QF—in Portuguese), the Peixe Bravo River Valley, and the Santo Antônio River Basin; in the state of Mato Grosso do Sul, the Morraria do Urucum; in the state of Pará, the Carajás Range (or *Serra dos Carajás*—SCJ—in Portuguese); and in the state of Bahia, the Caetité.

According to Madeira et al. (2015), three of these systems are entirely unprotected (the Peixe Bravo River Valley, the Caetité, and the Morraria do Urucum), while the other three (the QF, the SCJ, and the Santo Antônio River Basin) are partially inserted in Conservation Units (*Unidades de Conservação*—UCs—in Portuguese[2]). The QF has a small portion inserted in the *Rola Moça Range State Park*—a *Full Protection* UC—and a significant part inside an Environmental Protection Area (*Área de Proteção Ambiental*—APA—in Portuguese) located in

[2]Note that throughout this text, the direct translation of the official Brazilian nomenclatures for the distinct types of protected areas is used.

the south metropolitan area of Belo Horizonte (APA *Sul RMBH*—Belo Horizonte is the state of Minas Gerais' capital). The *Espinhaço Range* (UNESCO-MAB Biosphere Reserve since 2005), also located in the state of Minas Gerais—where the Santo Antônio River arises—is partially inserted in the *Picão River Municipal* APA (Pilar Hills), and most of the SCJ is inside the Carajás National Forest (*Floresta Nacional dos Carajás*—Flona Carajás—in Portuguese).

Parts of the QF and the SCJ are located inside mining concession areas (Fernandes et al. 2014; Carmo 2010), where many million tonnes of iron ore are annually taken from the mountains. This completely mischaracterizes environments that serve as habitats for many organisms (Milagres e Gomes et al. 2015). Moreover, some of these UCs, such as the *Rola Moça Range State Park* and the *Gandarela Waters National Park* (both located in the state of Minas Gerais), had their limits completely cropped to meet large mining companies' interests, disregarding both biological and geomorphological attributes that are essential to local communities demands and to the conservation of important species (Milagres e Gomes et al. 2015). On the other hand, the *Rola Moça Range State Park* is one of the rare *Full Protection* UCs that currently prioritizes the preservation of ecosystems located in ferruginous areas (Jacobi et al. 2015).

5.4 *Cangas*

Canga crusts comprise surface formations that result from evolutionary processes that occur both in the landscape and in the relief (Schaefer et al. 2015). These processes include the transport, deposition, chemical and mechanical degradation, weathering—which involves hydration, dissolution, hydrolysis—and polycyclic pedogenesis, as well as Fe, Al, and other trace elements supergene enrichment (Dorr 1964, 1969; McFarlane 1976; Meyer 1997; Schaefer et al. 2015; Souza and Carmo 2015). *Cangas* have a strong control on the vegetation distribution and are composed of fragments from surrounding rocks, cemented by iron (hydr)oxides (such as goethite, limonite, and/or hematite). They may, in most of the cases, be classified as soils derived from subaerial modifications of IFs (Schaefer et al. 2008, 2015; Souza and Carmo 2015).

Cangas usually occur, with some exceptions, at higher portions of the relief, forming plateaus interconnected with valleys and escarpments. They are found in tropical regions, such as Brazil, Australia, India, Guinea, Burkina Faso, Central African Republic, and Venezuela (Dorr 1973; Beauvais and Tardy 1991; Brown et al. 1994; Bekker et al. 2010; Monteiro et al. 2014; Souza and Carmo 2015). The iron extraction promotes a complete scenery change, since the entire *canga* sheet overlying the ore is completely removed and discarded to allow access to the deposits (Jacobi and Carmo 2008; Ferreira et al. 2015). Thus, the most visible impacts (that surpass the restricted areas of the enterprises' facilities) are the loss and degradation of both the local and the regional biodiversity (Ferreira et al. 2015; Souza and Carmo 2015).

5.5 The FGSs and the Iron Mining in Brazil

It is well known that the iron mining industry is of great economic importance to Brazil. The country's exploitable reserves have an average iron grade of 50.3% and accounted for more than 10% of the world's reserves in the past years (DNPM 2013; Ruchkys 2015), putting the Brazilian iron ore production frequently among the three largest in the world (DNPM 2014; Souza and Carmo 2015). The diversity of values (e.g., biological, archaeological, geological, speleological, hydrogeological, economic, functional, cultural, aesthetic, scientific, educational, among others, see references in Ruchkys 2015) generates disputes and conflicting interests in regions where iron occurs (Ruchkys and Machado 2015): on one hand, there is the need to preserve the geodiversity, and on the other hand, there are many interests linked to the use and economic exploitation of iron, since this element is a base material for the industrial development (Ruchkys 2015).

In 2015, most mineral commodities showed falling prices (DNPM 2015a), influenced by a lower global demand (which in turn, was related to (i) the transition of China's economic model, from a commodity-intensive economy to a service- and consumption-driven economy, (ii) a weak demand from emerging markets, and (iii) a global excess of mineral supplies—DNPM 2015a). Despite consecutive decreases in the Brazilian industrial production (of 8.3 and 9.1%, from January to December 2015 and from January to June 2016, compared to the same periods of 2014 and 2015, DNPM 2015b, 2016a), industrial activities related to mineral extraction showed an increase of 3.9% in 2015 (motivated mainly by an expansion in the production of some items, such as crude oil and iron ore, DNPM 2015b). On the other hand, during the first semester of 2016, the Brazilian mineral extraction industry, largely influenced by a reduction in the production of iron ore, decreased 14% (DNPM 2016a).

The states of Pará and Minas Gerais (where the SCJ and the QF are, respectively, located) were responsible, primarily due to their iron ore production, for the collection of 77.4% of all the country's mineral royalties during the first six months of 2016 (DNPM 2016a). In 2016, the state of Pará had an iron ore production that was over 147 million tonnes (an increase of a little more than 15% compared to 2015), and the exportation of this good was 16.3% greater than in 2015, making it the most exported commodity from this state that year (DNPM 2016b).

In 2015, the increase in the amount of iron ore exported was similar, in percentage, to the slump verified in the average price, also in terms of percentage (DNPM 2015b). In the first semester of 2016, once again, there was a 23.2% decrease in the exported value of iron (in dollars), whereas the amount of iron exported raised 5.7% (DNPM 2016a). Given that other mineral substances showed negative variations in their export values, the increase in the amount of iron ore exported greatly contributed to a less expressive trade balance deterioration of the Brazilian mining industry (in both 2015 and 2016), and also to a reconfiguration of the relative export participations of other mineral substances (DNPM 2015b, 2016a). This scenario highlights the importance of the mineral sector to the country's international commerce and the relevance of this element for the Brazilian export mining industry, despite price drops in the global market iron average price.

Although the FGSs represent extremely limited and isolated areas of the country, the iron mining exerts strong anthropic pressures over these systems (Jacobi et al. 2007). This puts several endemic species populations (which are threatened with extinction or show a very restricted distribution) at risk (Nunes 2009; Lopes et al. 2010; Vasconcelos and Rodrigues 2010). Moreover, as previously mentioned, the fauna and flora knowledge of these areas is very fragmented and incipient (Vasconcelos and Hoffmann 2015). Thus, the main threat to these ecosystems is the economic exploitation developed, as a rule, without proper environmental remediation (Jacobi et al. 2007; Takahasi 2015). Given their natural geographical isolation and their specific characteristics, the SCJ and the QF FGSs host unique ecosystems. As they are also the main targets for iron exploration in Brazil (Carmo 2010; Milagres e Gomes et al. 2015), they will be further discussed in more detail.

5.6 Quadrilátero Ferrífero (QF)

The QF is a mountainous region that represents the southern portion of the Espinhaço Range (Schaefer et al. 2015). It covers an area of approximately 7,000 km^2 (Roeser and Roeser 2010; Ruchkys 2015), and the ferruginous rock types are from the Archean, Proterozoic, and Cenozoic ages (Ruchkys 2015). The QF IFs belong to the Minas Supergroup, Itabira Group, Cauê Formation (Paleoproterozoic—Souza and Carmo 2015). In 2008, there were about 50 mineral exploitation mines in the QF region, covering an area of 2000 ha (Jacobi and Carmo 2008; Ferreira et al. 2015). Currently, several open-pit iron mines are active in the region and the mineralization is mainly associated with itabirites (Ruchkys 2015).

Studies about mutualism and parasitism in the QF region show that the *cangas* there, besides harboring a variety of microenvironments (Jacobi et al. 2007), are regionally distinct. Some plants are particularly important due to their interaction with other species in several trophic levels. Also, the large number of mountains in the state of Minas Gerais often act as dispersal barriers for the wildlife, making each vegetation patch in the landscape unique (Jacobi et al. 2015).

Nowadays, there are only a few well-preserved natural areas in the QF, since a great portion of the region is owned by mining companies and, therefore, is affected by the iron ore mining, which causes an irreversible loss of the *cangas* (Jacobi et al. 2011; Messias and Carmo 2015). The *cangas* in the QF show high diversity and endemism and are subject to the mining pressure, the urban expansion of metropolitan areas, the illegal hunting, the improper flora management, deforestation, and finally, natural or human-generated fires (Jacobi et al. 2007; Gibson et al. 2010; see references in Salgado-Labouriau and Ferraz-Vicentini 1994). These activities usually contribute to impair the connectivity between the fragments and to island the biodiversity (Morcatty et al. 2013; Takahasi 2015; Milagres e Gomes et al. 2015).

Carmo and Jacobi (2013) reported the existence of 102 families, 418 *genera*, and 1,080 species of vascular plants in the QF region (Messias and Carmo 2015). At least 89 plant species with high conservation values, many of them rare and endangered, occur solely in these *canga* areas (Carmo 2010, Carmo et al. 2012; Ruchkys 2015). The QF region records six bird species listed as endangered at global, national, and local levels (Lopes et al. 2010; Vasconcelos and Hoffmann 2015). In addition, 26 flora species are mentioned in official endangered lists in Brazil (Brasil 2014) and in the state of Minas Gerais (COPAM 1997), some of them endemic to the QF *cangas*. However, other QF endemic species, whose populations are also declining due to the loss and degradation of natural FGSs areas, are not included in official lists, and several *canga* spaces in that region are out of *Full Protection* UCs (Messias and Carmo 2015).

5.7 Serra dos Carajás (SCJ)

The SCJ mineral province occupies part of the states of Pará, Maranhão, and Tocantins in Brazil (Santos 1986; Mota et al. 2015). It is located in the southeastern portion of the Amazonian Craton and is characterized by the occurrence of lithologies associated with the Precambrian Grão Pará Group, Carajás Formation (Pereira 2009; Souza and Carmo 2015). In addition to the peculiarity of its vegetation, the SCJ is considered one of the planet's most expressive mineral areas and the richest in Brazil (Freitas 1986; Santos 1986). The extractive processes, which are related to iron, copper, nickel, gold, manganese, sand, gneiss quarries, and amethyst, among others, give an idea of the dimension and the variety of rocks and, by correlation, the diversity of plant communities (Mota et al. 2015).

The region hosts a total of five UCs (Madeira et al. 2015), and the SCJ mountains are almost entirely within the Flona Carajás boundaries, a *Sustainable Use* UC that covers over 411,000 ha (Decree 2486/1998—Brasil 1998) and is the most important preservation area of the SCJ (Madeira et al. 2015). The highest hills in the Flona Carajás consist of IFs that form a dense area of *cangas* (Mota et al. 2015) that cover less than 12,000 ha. About 30% of these *canga* areas are already interfered by mining activities or are in licensing processes, what leaves only about 8,000 ha intact (Madeira et al. 2015).

The Flona Carajás was officially created in February 1998, a hectic time in the national political scene, when there were major demonstrations against the privatization of the *Companhia Vale do Rio Doce* (CVRD, the now privatized Vale, Madeira et al. 2015). Unlike other *Flonas*, its creation decree guaranteed the CVRD mining activities among the UC's objectives. It was then clear the strategy to, on one hand, use the main biodiversity conservation policy instrument (i.e., the UCs) to provide and ensure a long-term mineral exploration, and on the other hand, use the abundance of mining funds to implement the UCs (Madeira et al. 2015).

According to Madeira et al. (2015), large and important high-grade iron ore deposits, with an average content of 65.4% (all associated with FGSs), stand out in the SCJ and are estimated at about 18 billion tonnes (http://www.mining-technology.com/projects/carajas/—accessed February 2018). The SCJ was discovered in 1967 by the United States Steel Company (US Steel), who formalized an agreement with the Brazilian Government in 1970, giving rise to the AMZA Company (Amazon Mining S.A.). The then CVRD held 51% of the share capital and the US Steel, 49%. In 1974, AMZA got the rights to operate over the entire area of Carajás (Santos 1986; Mota et al. 2015).

For Santos (2010), the Carajás UCs were designed to create a shield area to keep mining activities away from human intrusions. Currently, according to Mota et al. (2015), the surroundings of the Flona Carajás host seven settlements related to the National Institute for Colonization and Agrarian Reform (*Instituto Nacional de Colonização e Reforma Agrária*—INCRA—in Portuguese), often criticized for allowing people with high incomes to receive lands destined for underprivileged people, who are waiting in line for over 10 years. Also, a large portion of the Flona Carajás is limited by the urban area of the Parauapebas municipality (Carvalho 2010). In this context, the peculiar vegetation over the *cangas* at the SCJ has been eliminated at a pace that threatens to run over the flora and fauna knowledge and the definition of strategies that may ensure their conservation (Mota et al. 2015).

A consistent compilation of endemic flora species that occur in the SCJ does not exist; however, a brief analysis of the information available (both in literature and in herbariums) shows there are more than 40 described species, presumably endemic, among the botanical material collected in that region (Mota et al. 2015). From a conservationist point of view, important Brazilian flora species are listed as threatened with extinction (Fundação Biodiversitas 2008) or at risk (Schaefer et al. 2015). Also, several intermittent watercourses (shallow lakes, which alternate with exposed and dry soils during summer) are formed at the Carajás ferruginous mountaintops during rainy periods (Mota et al. 2015). In these drastic water availability alternation environments, annual communities, which can vary the floristic composition between locations within the same area and even throughout the annual cycle, settle every season. Some existing lagoons (e.g., the ones in a mountain chain named *South Hills*, located inside the Flona Carajás) keep their water column throughout the whole annual cycle, even if their volume is drastically reduced during the dry season. However, the real scenario is that at least two perennial lagoons located at the *North Hills* mountain chain were completely eliminated by mining activities in the past (Mota et al. 2015) and that might as well happen in the future.

The environmental protection and the full demarcation or ratification of indigenous territories should have been first priorities for the SCJ, as suggested by Treece (1987) more than 30 years ago. Unfortunately, it is clear the *emergency* and *long-term* programs this author envisioned for that region never came out of the paper. As pointed out by Treece (1987), "tribal peoples can only freely decide their future in the certain knowledge that the forests, lands, and rivers, the basis of their economic existence and cultural identity, will still be there tomorrow". Today, this is an uncertainty not only for the indigenous peoples in Brazil, but for all present and future generations, as the Amazonian rivers represent almost 20% of all the planet's freshwater supply.

5.8 The Relation Between Mining, Hydrogeology, and FGSs in Brazil

The *cangas* high porosity and permeability rates are fundamental to hydrogeological processes (Silva 2001). It is estimated that together, only the SCJ and the QF regions shelter thousands of cavities (Carmo and Kamino 2015b; Tobias Junior et al. 2015). The association between aquifers, FGSs, and water security is evident, as well as the undeniable irreversibility of mining-related damage to water reserves (Madeira et al. 2015) that are strategic and provide vital ecosystem services to society on a daily basis.

The *canga* zones of the QF FGSs are important recharge areas for major aquifers hosted by the Cauê Formation, which is mostly composed of BIFs (Carmo and Kamino 2015b). There are thousands of springs in this hydrogeological unit, some of them with flows that can reach up to 500 m^3/h (Beato et al. 2006). The Cauê Formation exhibits great textural and compositional variability and hosts, besides the iron ore, the main QF aquifers (Gama and Matias 2015). Although in subsurface, underground water may be affected by pollution, and if it is more difficult to contaminate it, the opposite is also true: Once polluted, it is very difficult to reverse the condition and the investment needed is often extremely expensive (Gama and Matias 2015).

Regarding the many adverse effects associated with the mining activity, the mineral concentration processes should be noted. They generate a significant amount of waste (i.e., remaining materials from the beneficiation processes and the metal concentration in industrial plants), whose granulometric characteristics depend on the raw ore being extracted and the beneficiation process itself. The deposition of this waste must be controlled and carried out according to both the rheological properties and the chemical composition of the materials (Gama and Matias 2015). For that purpose, there are the containment dams, also known as tailings dams.

Whatever the function of these dams, their design must fulfill the requirements needed throughout the lifetime of the mine (Gama and Matias 2015), otherwise environmental disasters, such as the two Samarco (joint venture between Vale and BHP Billiton, the world's largest mining company) dams that burst in the municipality of Mariana (state of Minas Gerais) on November 05, 2015, will likely happen. Following this accident, nineteen people lost their lives and between 30 and 62 million m^3 of mud (i.e., tailings) from the *Fundão* and the *Santarém* dams spread, reaching the Atlantic Ocean on November 21, 2015, in the state of Espírito Santo. Almost three years later, many families still wait for any kind of help from Samarco, Vale, or BHP Billington, but none of them is willing to assume any responsibility. Important to notice that the same Samarco joint venture owns and operates the *Samarco Alegria* open-pit iron mine, also located in the state of Minas Gerais and considered the world's second largest iron mine, estimated to hold 1.89 billion tonnes of proven 40.2% grade iron ore reserves (http://www.mining-technology.com/features/featurethe-worlds-11-biggest-iron-ore-mines-4180663/—accessed February 2018).

The burst of tailing dams is not something new to Brazil, though. In 1987, a dam from the *Pitinga* mine burst in the indigenous *Waimiri-Atroari* territory, located in the state of Amazonas, North Brazil. The indians water supply was contaminated with mud and toxic waste that drove away the fish and caused untold disruption to the hydrology of the region (Treece 1987). This is just one among the many cases of indigenous peoples (e.g., the *Catetê*, the *Parakanã*, the *Anambé*, the *Guamá*, the *Apinayé*, the *Awa-Gurupi*, the *Krikati,* and others; see more in Treece 1987) that were, and still are, negatively affected by mining operations in Brazil.

5.9 Regulatory Inconsistencies

In Brazil, hydrogeological studies for specific purposes (e.g., lowering the water table, dams for power generation, flood control) are subject to specific legislation and responsibility of different institutions: In the Federal level, it is the recently created National Mining Agency (*Agência Nacional de Mineração*—ANM—in Portuguese, the former National Mineral Production Department or *Departamento Nacional de Produção Mineral*—DNPM—in Portuguese), and in the case of the Brazilian states, other public entities (Gama and Matias 2015). For instance, the public agent in charge in the state of Minas Gerais is the Minas Gerais Institute for Water Management (*Instituto Mineiro de Gestão das Águas*—IGAM—in Portuguese). All actions performed by these agencies should be coordinated based on the National Water Resources Policy (Law 9433/97, Brasil 1997, *Política Nacional de Recursos Hídricos* —PNRH—in Portuguese); however, there is little attention from the government regarding the protection of the country's watercourses (Gama and Matias 2015).

In this context, the 1988 Brazilian Federal Constitution (Constituição 1988, Brasil 1988) is quite contradictory in some definitions, making it difficult to find the much-needed institutional alignment that may lead to reasonable approaches to the many problems. The Constituição 1988 provides that are the Union (i.e., the Brazil State) properties: *lakes, rivers, and any watercourses within its domain, or that wash more than one state; that serve as boundaries with other countries; or that extend into foreign territory or from them proceed* (Article 20, Section III), and also *mineral resources, including those from subsurface* (Article 20, Section IX); on the other hand, it says that are the states (not the Union, or the Brazil State) properties *surface or groundwaters, flowing, emerging or in deposit* (Article 26, Section I). Hence, it can be concluded that even if they have limits that spread over more than one Brazilian state, and although they have a subsurface origin, groundwaters are not properties of the Union (the Brazil State). This makes it very hard to integrate the management of water resources—although this is precisely one of the three objectives of the PNRH, which is the main existing instrument able to regulate the water consumption and prevent the rationing in Brazil.

Indeed, each Brazilian state has the will to decide about the aquifers' fate and exploitation in their territory. Unfortunately, this power has led to severe consequences and manifestly wrong decisions, such as the one made by the state of São

Paulo (southeastern Brazil) government: throughout 2015, the water distribution from the Cantareira System to the population operated in the first *dead volume*, and even in the second *dead volume* quota for some months. Only on December 30, 2015, not thanks to its government, the state of São Paulo had higher than expected rainfalls, which allowed the Cantareira System to finally return its operations back to the useful volume, after 17 months.

Water is a *public good* in the PNRH, therefore subject to issuance by the National Water Agency (*Agência Nacional de Águas*—ANA—in Portuguese) when classified as a Union (Brazil State) property, and by the states' water agencies when considered a state property. In the National Environmental Policy (Law 6938/1981, Brasil 1981), water is a *diffuse good*, since its Article 3, Section V, lists *the atmosphere, inland waters, surface and groundwater, estuaries, territorial sea, soil, the subsoil, the biosphere elements, the fauna, and the flora* as *environmental resources* in a general sense. Last, but not least, there is also the concept of *mineral water* in Brazilian laws (those from *natural sources, or from artificially captured sources, possessing, when compared to ordinary waters, distinct chemical composition, physical or physicochemical properties,* and that have *characteristics that give them a medicamentous action* (Mineral Waters Code, Decree 7841/1945, Brasil 1945). In the latter case (i.e., mineral water), the issuance must be given by the ANM (former DNPM), not the ANA. Again, the many contradictory and confusing definitions tend to widely impair the life of (human, fauna, and flora) populations. In this context, where the management of water resources is a key factor to ensure fundamental trans-specific rights, the FGSs conservation is nothing but critical.

Regarding the seriousness of the threats imposed over the availability of water resources and the guarantee of water multiple uses in Brazil (Law 9433/97, PNRH, Brasil, 1997), it is crucial to highlight the country's problems associated with deficits and inefficiencies involving the water treatment, the sanitation implementation, and the environmental management of solid waste. The direct release of untreated sewage into rivers, lakes, and oceans, as well as landfills and garbage dumps in cities, eventually ends up contaminating, respectively, surface and groundwater (Gama and Matias 2015). In Brazil, 82.5% of the population is served with treated water. However, 37% of the collected and treated water is simply lost (e.g., leaks, theft and illegal connections, lack of measurement or incorrect measurements in the water consumption). In addition, 48.6% of the population has no access to sewage collection and only 40% of the country's sewage is treated: Only 10 of the 100 largest Brazilian cities treat over 80% of its wastewater (Brasil 2016). This complex scenario asks, urgently, for a consistent protection of environmental assets, including the ones associated with FGSs.

5.10 Risks of Retreats to Environmental Conservation

The implementation of major infrastructure projects, the disordered urbanization, and the occupation of coastal and sloped areas, as well as the predatory and unplanned mining activities, are some of the many threats to the geological heritage in a broad sense and, specifically, to the heritage associated with FGSs (Ruchkys 2015). Added to these factors is the problem, very common in many countries, of rare minerals and fossils piracy or destruction. The geological heritage is not renewable; once destroyed, it does not regenerate and a part of the planet's memory is simply lost forever (Ruchkys 2015).

According to Carmo et al. (2012), areas officially granted for iron mining companies cover about 300,000 km^2 from Brazil. Inside these areas, lie more than 99% of the *canga* outcrops (Ruchkys 2015). As previously stated, despite the FGSs economic and environmental relevance and the growing number of publications about the biology found in iron ore areas, very little is known about the associated flora and fauna, and much less on the ecological roles played by the organisms that occur in these regions (Milagres e Gomes et al. 2015). A considerable part of these environments is, in both medium and long terms, at risk of disappearing due to anthropogenic pressures (Jacobi et al. 2007, Tavares et al. 2012; Carmo 2010; Fernandes et al. 2014; Milagres e Gomes et al. 2015).

The inevitable coincidence between the existence of *cangas* and iron ore explains why almost no protection is given to the biota associated with FGSs inside UCs (Madeira et al. 2015). Nevertheless, the high endemism and the uniqueness of ecological and evolutionary processes (widely discussed in the book *Ferruginous Geosystems in Brazil*, Carmo and Kamino 2015a) make these areas very important conservation targets. The wide and direct impact addressed to *canga* regions and, consequently, to groundwater recharge areas, must be understood and evaluated regarding the latest national and international commitments for biodiversity conservation (Madeira et al. 2015).

Given the FGSs' contribution to evolutionary processes, the wealth of endemic species and the persistence of some specific features (e.g., underground natural cavities), it is appropriate to presume that at least a portion of *canga* areas is preserved inside each of the six main Brazil FGSs (Madeira et al. 2015). In the SCJ, there are negotiations since 2004 about the protection of a *Minimum Canga Area* that should represent a percentage of what was originally there and should be technically understood as sufficient to prevent biodiversity losses, as well as to ensure evolutionary processes. Unfortunately, the QF area has been almost entirely exploited without, until recently, any concern with the *cangas* conservation (Madeira et al. 2015). Moreover, the pressure on the Espinhaço Range (region in the state of Minas Gerais that contains one of the richest floras in Brazil, hosts a great number of endemic species, and is the most recent iron mining target in the country—Echternacht et al. 2011) has caused huge environmental and social impacts due to the construction of pipelines in areas that have no ore at all (e.g., Conceição do Mato Dentro region, also in the state of Minas Gerais, Madeira et al. 2015).

Brazil has already had two National Parks, called *Paulo Afonso* and *Sete Quedas*, entirely vanished in the 1960s and 70s to make way for mega-dams (Madeira et al. 2015), even though it is well known that the larger the dam the less reliable it is and the worse the environmental problems it will likely cause (Treece 1987). It is expected that environmental regulations can stop situations like the one that happened in 2012, when there was a reduction of several protected areas in order to allow the exploitation of Amazonian rivers hydroelectric potential. In this context, the licensing process is the final chance to prevent the mining advance on important reminiscent areas (Madeira et al. 2015), especially if they are not inside any UCs (what is not unlikely, given the country's current political and economic situation and its dependence on commodities exportation). However, on April 2016, the Commission of Constitution, Justice and Citizenship (*Comissão de Constituição, Justiça e Cidadania*—CCJ—in Portuguese) from the Federal Senate approved an Amendment to the Constitution Proposal (in Portuguese, *Proposta de Emenda Constitucional*—PEC—number 65/2012) that may definitely (and very surprisingly) end the need of environmental licensing for constructions throughout the country. If this PEC is ultimately approved by the Brazilian National Congress, formed by both senators and congressmen, the simple presentation of the Environmental Impact Study (*Estudo de Impacto Ambiental*—EIA—in Portuguese) implies authorization to execute the venture, which can no longer be, later on, suspended or cancelled.

The management of the Brazilian water resources—especially regarding the ones related to FGSs—requires specific and integrated actions that include educational activities aiming to encourage the environmental preservation, as well as the conscious and responsible use of water (Gama and Matias 2015). The environmental licensing should be increasingly careful and not shaped to satisfy private or large companies' interests as it has been in Brazil over the years (Madeira et al. 2015). In any case, if the Brazilian Government does not stop to promote policies that clearly favor conflicting interests regarding the environment and the observance of trans-specific rights (that ensure a balanced environment for present and future generations), a particularly tragic and disastrous adversity drawn by the (soon to be official) Anthropocene (Crutzen 2002; Danowski and Viveiros de Castro 2014) will be experienced, at all scales, as a fact and as an irremediable experience. As pointed out by Treece (1987), wide climatic repercussions become more credible each day, as the knowledge about the functioning of forests grows.

5.11 Final Remarks

Since the global financial crisis in 2008 and 2009, and under a neo-developmentalist project designed as an alternative to the previous neoliberal policies, the Brazilian economy has experienced significant growth for years. The Brazil's Growth Acceleration Program (*Programa de Aceleração do Crescimento*—PAC—in Portuguese), which gained special projection (in both its versions or phases—2007 and 2011), had the goal to keep the national economy active through strategic investments in structuring sectors, particularly the ones associated with the country's construction field: social, urban, logistics, and energetic infrastructure. In this sense, one of the latter investment consequences was a disastrous venture for the *Belo Monte Hydroelectric Plant* construction, in the state of Pará (northern Brazil), a true environmental catastrophe that goes on until today.

Whatever the result of the uncertain Brazilian political scene, there is no doubt that, either in the neoliberal or the neo-developmentalist line, strong political pressures over the environmental agenda will prevail. Regardless of which political force shall drive the national economic policy, everything indicates that the environmental plans, the indigenous peoples, and the traditional communities will be sacrificed with the excuse to recover the economic growth. Tendentiously, social and collective rights, indigenous lands, *quilombola*[3] territories, federal and state's UCs, and all the environmental legislation, including, in particular, the laws related to the licensing of ventures that cause significant environmental impacts, will likely be tensioned. Above all, the fate of Brazilian FGSs, these complex formations of immeasurable importance, will continue to be at risk.

In this scenario, the geology of Brazil is, at this very moment, ethically challenged regarding the role it will play on its practices and professional applications. This choice resonates at the political, institutional, academic, and individual prisms. The story that follows is yet to be written, but deserves the most serious consideration in the different scopes in which the geology exercise is made in the country. It is proposed in this paper that the geology community leans itself toward a plural, trans-specific, and anti-utilitarian geoethic; that it denies the instrumental appropriation related to market pressures. Professionals acting in this area, as well as other Earth Science professionals, should adopt a geoethic aligned to the acknowledgment that all forms of life hold an unequivocal right to exist. And regarding the FGSs in Brazil, it reclaims that geologists and related experts show an uncompromised defense for the conservation of these unique ecologic, biologic, and geologically formations.

Acknowledgements The author would like to thank one anonymous reviewer, Drs. José Eloi Guimarães Campos, Tiago Eli de Lima Passos, and Celeida Santos for their considerations.

[3]*Quilombolas* are the inhabitants of *Quilombos*, term that has "new significances in the specialized literature, also for groups, individuals, and organizations. Besides having a historical content, it has been resemantized to designate the present situation of black segments in regions and contexts in Brazil. *Quilombo* does not refer to archaeological residuals or remnants of a temporal occupation or a biological comprovation. It is not about isolated groups or strictly homogeneous populations. They were not always constituted from insurrection or rebel movements. Above all, they consist in groups that developed daily resistance practices regarding both the maintenance and reproduction of characteristic ways of life and the consolidation of their own territory. The identity of these groups is not defined by their size or number, but by their lived experience and the shared versions of their common trajectories and their continuity as a group. They constitute ethnical groups regarded by the Anthropology as organizational type that grants membership by standards and means of affiliation or exclusion" (ABA 1994).

References

ABA—Associação Brasileira de Antropologia (1994) Documento do Grupo de Trabalho sobre Comunidades Negras Rurais. Available for full download at the Associação Brasileira de Antropologia (ABA) website: http://www.abant.org.br/conteudo/005COMISSOESGTS/quilombos/DocQuilombosABA_1a.pdf. Accessed 07 Feb 2018

Beato DAC, Monsores ALM, Bertachini AC (2006) Potencial aquífero nos metassedimentos do Quadrilátero Ferrífero - região da APA SUL RMBH - MG. In: XIV Congresso Brasileiro de Geologia, pp. 1–20. Available at the Associação Brasileira de Águas Subterrâneas (ABAS) website: https://aguassubterraneas.abas.org/asubterraneas/article/view/23106/15221. Accessed 07 Feb 2018

Beauvais A, Tardy Y (1991) Formation et degradation des cuirasses ferrugineuses sous climat tropical humide, à la lisière de la forêt équatoriale. In: Comptes Rendus de l'Académie des Sciences, Série II, Gauthier-Villars, Paris, France, t. 313(13):1539–1545

Bekker A, Slack JF, Planavsky N, Krapež B, Hofmann A, Konhauser KO, Rouxel OJ (2010) Iron formation: the sedimentary product of a complex interplay among mantle, tectonic, oceanic, and biospheric processes. Econ Geol 105(3):467–508

Brasil (1945) Decreto-Lei 7.841, Código de Águas Minerais, de 08 de agosto

Brasil (1981) Lei nº 6.938, Política Nacional do Meio Ambiente, de 31 de agosto

Brasil (1988) Constituição Federal

Brasil (1997) Lei nº 9.433, Política Nacional de Recursos Hídricos, de 08 de janeiro

Brasil (1998) Decreto 2.486, Criação da Floresta Nacional de Carajás, no Estado do Pará, de 02 de fevereiro

Brasil (2014) Ministério do Meio Ambiente (MMA). Portaria n. 443, de 17 de dezembro. Lista Nacional Oficial de Espécies da Flora Ameaçadas de Extinção. Diário Oficial [da República Federativa do Brasil], Brasília, DF, n. 245, 18 dez. 2014. Seção 1. pp 110–121. Available at the Seleção de Matérias de Interesse da Justiça Eleitoral website: http://sintse.tse.jus.br/documentos/2014/Dez/18/portaria-no-443-de-17-de-dezembro-de-2014. Accessed 07 Feb 2018

Brasil (2016) Ministério das Cidades. Secretaria Nacional de Saneamento Ambiental – SNSA. Sistema Nacional de Informações sobre Saneamento (SNIS): *Diagnóstico dos Serviços de Água e Esgotos—2014*. Brasília: SNSA/MCIDADES. 212 pp. Available at the Sistema Nacional de Informações sobre Saneamento website: http://www.snis.gov.br/diagnostico-agua-e-esgotos/diagnostico-ae-2014. Accessed 07 Feb 2018

Brown ET, Bourlès DL, Colin F, Sanfo Z, Raisbeck GM, Yiou F (1994) The development of iron crust lateritic systems in Burkina Faso, West Africa examined with in-situ-produced cosmogenic nuclides. Earth Planet Sci Lett 124(1–4):19–33

Carmo FF (2010) Importância ambiental e estado de conservação dos ecossistemas de cangas no Quadrilátero Ferrífero e proposta de Áreas-Alvo para a Investigação e Proteção da Biodiversidade em Minas Gerais. Dissertação de Mestrado. Belo Horizonte: Universidade Federal de Minas Gerais (UFMG), Instituto de Ciências Biológicas (ICB), Programa de Pós-graduação em Ecologia, Conservação e Manejo da Vida Silvestre (ECMVS), 90 pp

Carmo FF, Jacobi CM (2013) A vegetação de canga no Quadrilátero Ferrífero, Minas Gerais: caracterização e contexto fitogeográfico. Rodriguésia 64(3):527–541

Carmo FF, Kamino LHY (2015a) Geossistemas Ferruginosos do Brasil: áreas prioritárias para conservação da diversidade geológica e biológica, patrimônio cultural e serviços ambientais. In: Carmo FF, Kamino LHY (Orgs.), Belo Horizonte, Brazil, 3i Editora, 1 ed., vol 1, 552 pp

Carmo FF, Kamino LHY (2015b) Introdução – Geossistemas Ferruginosos. In: Geossistemas Ferruginosos do Brasil: áreas prioritárias para conservação da diversidade geológica e biológica, patrimônio cultural e serviços ambientais. Belo Horizonte, Brazil, 3i Editora, 1 ed., vol 1, pp 23–46

Carmo FF, Campos IC, Jacobi CM (2012) Cangas: ilhas de ferro estratégicas para a conservação. Ciência Hoje 295(50):48–53

Carvalho AS (2010) Mamíferos de médio e grande porte na Floresta Nacional de Carajás, Pará: riqueza, abundância e efeitos da fitofisionomia e do impacto da mineração. Dissertação de Mestrado. Rio de Janeiro, Universidade do Estado do Rio de Janeiro (UERJ), 133 pp

COPAM (1997) Lista das espécies ameaçadas de extinção da flora do Estado de Minas Gerais. Deliberação COPAM 085/97. Available at the Fundação Biodiversitas website: http://www.biodiversitas.org.br/florabr/MG-especies-ameacadas.pdf. Accessed 07 Feb 2018

Cornell RM, Schwertmann U (2003) The iron oxides: structure, properties, reactions, occurrences and uses, 2nd edn. Wiley-VCH Verlag GmbH and Co. KGaA, Weinheim, Germany, p 703

Crutzen P (2002) Geology of mankind. Nature 415:23

Curie C, Briat JF (2003) Iron transport and signaling in plants. Annu Rev Plant Biol 54:183–206

Danowski D, Viveiros de Castro E (2014) Há mundo por vir? Ensaio sobre os medos e os fins. Cultura e Barbárie Editora and Instituto Socioambiental (ISA). Desterro, Florianópolis, Brazil, 176 pp

DNPM—Departamento Nacional de Produção Mineral (2013) Sumário Mineral volume 33. Coordenador: Lima, T. M. Brasília, DNPM, 148 pp

DNPM—Departamento Nacional de Produção Mineral (2014) Sumário Mineral volume 34. Coordenador: Lima, T. M. Brasília, DNPM. DNPM, 152 pp

DNPM—Departamento Nacional de Produção Mineral (2015a). Informe Mineral 1/2015 (janeiro a junho). Brasília, DNPM, 23 pp

DNPM—Departamento Nacional de Produção Mineral (2015b) Informe Mineral 2/2015 (julho a dezembro). Brasília, DNPM, 22 pp

DNPM—Departamento Nacional de Produção Mineral (2016a) Informe Mineral 1/2016 (janeiro a junho). Brasília, DNPM, 15 pp

DNPM—Departamento Nacional de Produção Mineral (2016b) Informe Mineral do Estado do Pará (ano base 2016). Brasília, DNPM, 10 pp

Dorr JVN (1964) Supergene iron ores of Minas Gerais, Brazil. Econ Geol 59(7):1203–1240

Dorr JVN (1969) Physiographic, stratigraphic and structural development of Quadrilátero Ferrífero, Minas Gerais, Brasil. In: Geological Survey Professional Paper 641-A, United States Geological Survey (USGS). United States Government Printing Office, Washington, 110 pp

Dorr JVN (1973) Iron-formation in South America. Econ Geol 68:1005–1022

Echternacht L, Trovó M, Oliveira CT, Pirani JR (2011) Areas of endemism in the Espinhaço range in Minas Gerais, Brazil. Flora—Morphol Distrib Funct Ecol Plants 206(9):782–791

Fernandes GW, Barbosa NPU, Negreiros D, Paglia AP (2014) Challenges for the conservation of vanishing megadiverse rupestrian grasslands. Natureza Conservação 12(2):162–165

Ferreira RL, de Oliveira MPA, Silva MSS (2015) Biodiversidade Subterrânea em Geossistemas Ferruginosos. In: Geossistemas Ferruginosos do Brasil: áreas prioritárias para conservação da diversidade geológica e biológica, patrimônio cultural e serviços ambientais. Belo Horizonte, Brazil, 3i Editora, 1 ed., vol 1, pp 195–231

Freitas MLD (1986) Algumas considerações sobre a região-programa. In: de Almeida JMG Jr. (Org.) Carajás: Desafio Político, Ecologia e Desenvolvimento. Brasília/CNPq; Brasiliense, São Paulo, pp 22–29

Fundação Biodiversitas (2008) Lista vermelha da flora ameaçada de extinção. Available at the Fundação Biodiversitas website: http://www.biodiversitas.org.br/florabr/lista_florabr.pdf. Accessed 07 Feb 2018

Gama EM, Matias JP (2015) Hidrogeologia e os Geossistemas Ferruginosos. In: Geossistemas Ferruginosos do Brasil: áreas prioritárias para conservação da diversidade geológica e biológica, patrimônio cultural e serviços ambientais. Belo Horizonte, Brazil, 3i Editora, 1 ed., vol 1, pp 103–124

Gibson N, Yates CJ, Dillon R (2010) Plant communities of the ironstone ranges of South Western Australia: hotspots for plant diversity and mineral deposits. Biodivers Conserv 19(14):3951–3962

Gibson N, Meissner R, Markey AS, Thompson WA (2012) Patterns of plant diversity in ironstone ranges in arid south western Australia. J Arid Environ 77:25–31

Gibson N, Coates D, van Leeuwen S, Yates C (2015) Hot, dry and ancient: banded iron formations of Western Australia. In: Geossistemas Ferruginosos do Brasil: áreas prioritárias para conservação da diversidade geológica e biológica, patrimônio cultural e serviços ambientais. Belo Horizonte, Brazil, 3i Editora, 1 ed., vol 1, pp 361–409

Jacobi CM, Carmo FF (2008) The contribution of ironstone outcrops to plant diversity in the Iron Quadrangle, a threatened Brazilian landscape. Ambio 37(4):324–326

Jacobi CM, Carmo FF (2011) Life-forms, pollination and seed dispersal syndromes in plant communities on ironstone outcrops, SE Brazil. Acta Bot Brasilica 25(2):395–412

Jacobi CM, Carmo FF, Vincent RC, Stehmann JR (2007) Plant communities on ironstone outcrops: a diverse and endangered Brazilian ecosystem. Biodivers Conserv 16(7):2185–2200

Jacobi CM, Carmo FF, Campos IC (2011) Soaring extinction threats to endemic plants in Brazilian metal-rich regions. Ambio 40(5):540–543

Jacobi CM, Mourão FA, Antonini Y, Fagundes R, Kumagai AF (2015) Interações Ecológicas em Ambientes de Canga do Quadrilátero Ferrífero, Minas Gerais. In: Geossistemas Ferruginosos do Brasil: áreas prioritárias para conservação da diversidade geológica e biológica, patrimônio cultural e serviços ambientais. Belo Horizonte, Brazil, 3i Editora, 1 ed., vol 1, pp 411–428

Lopes LE, Malacco GB, Alteff EF, Vasconcelos MF, Hoffmann D, Silveira LF (2010) Range extensions and conservation of some threatened or little known Brazilian grassland birds. Bird Conserv Int 20(1):84–94. Also available as Lopes et al., 2009, 19:1–11

Madeira JA, Martins FD, Ribeiro KT, Carvalho AS (2015) Geossistemas Ferruginosos e Áreas Protegidas. In: Geossistemas Ferruginosos do Brasil: áreas prioritárias para conservação da diversidade geológica e biológica, patrimônio cultural e serviços ambientais. Belo Horizonte, Brazil, 3i Editora, 1 ed., vol 1, pp 521–543

McFarlane MJ (1976) Laterite and landscape. Academic Press Inc., London, p 164

Messias MCTB, Carmo FF (2015) Flora e Vegetação em Substratos Ferruginosos do Sudeste do Quadrilátero Ferrífero. In: Geossistemas Ferruginosos do Brasil: áreas prioritárias para conservação da diversidade geológica e biológica, patrimônio cultural e serviços ambientais. Belo Horizonte, Brazil, 3i Editora, 1 ed., vol 1, pp 335–360

Meyer R (1997) Paleolaterites and paleosols: imprints of terrestrial processes in sedimentary rocks. In: Balkema AA. Rotterdam, 151 pp

Milagres e Gomes A, Dias da Silva LH, da Cunha Tavares V (2015) Morcegos Associados à Ecossistemas Ferruginosos de Minas Gerais. In: Geossistemas Ferruginosos do Brasil: áreas prioritárias para conservação da diversidade geológica e biológica, patrimônio cultural e serviços ambientais. Belo Horizonte, Brazil, 3i Editora, 1 ed., vol 1, pp 233–258

Monteiro HS, Vasconcelos PM, Farley KA, Spier CA, Mello CL (2014) U-Th/He geochronology of goethite and the origin and evolution of cangas. Geochim Cosmochim Acta 131:267–289

Morcatty TQ, El Bizri HR, Carneiro HCS, Biasizzo RL, Alméri CRO, Silva ES, Rodrigues FHG, Figueira JEC (2013) Habitat loss and mammalian extinction patterns: are the reserves in the Quadrilátero Ferrífero, southeastern Brazil, effective in conserving mammals? Ecol Res 28 (6):935–947

Mota NFO, Silva, LVC, Martins FD, Viana PL (2015) Vegetação sobre sistemas ferruginosos da Serra dos Carajás. In: Geossistemas Ferruginosos do Brasil: áreas prioritárias para conservação da diversidade geológica e biológica, patrimônio cultural e serviços ambientais. Belo Horizonte, Brazil, 3i Editora, 1 ed., vol 1, pp 289–315

Nielsen JM (1960) The radiochemistry of iron. U. S. Atomic Energy Commission, Subcommittee on Radiochemistry, National Academy of Sciences/National Research Council. Nuclear Science Series, 48 pp

Nunes AP (2009) Campainha-azul (Porphyrospiza caerulescens) (Passeriformes: Emberizidae) na Bacia do Alto Paraguai, Brasil: distribuição e conservação. Atualidades Ornitológicas, 152: 5–7

Pepper M, Doughty P, Keogh JS (2013) Geodiversity and endemism in the iconic Australian Pilbara region: a review of landscape evolution and biotic response in an ancient refugium. J Biogeogr 40(7):1225–1239

Pereira RMP (2009) Geologia da Região Sul da Serra Norte e Características do Minério de Ferro do Depósito N8, Província Mineral Carajás. Dissertação (Mestrado em Geologia). Universidade Federal de Minas Gerais (UFMG), Belo Horizonte, 143 pp

Roeser HMP, Roeser PA (2010) O Quadrilátero Ferrífero - MG, Brasil: aspectos sobre sua história, seus recursos minerais e problemas ambientais relacionados. Geonomos 18(1):33–37

Rosière CA, Chemale F Jr (2000) Itabiritos e minérios de ferro de alto teor do Quadrilátero Ferrífero – uma visão geral e discussão. Geonomos 8(2):27–43

Ruchkys UA (2015) Sítios Geológicos e Propostas Brasileiras de Geoparques em Geossistemas Ferruginosos. In: Geossistemas Ferruginosos do Brasil: áreas prioritárias para conservação da diversidade geológica e biológica, patrimônio cultural e serviços ambientais. Belo Horizonte, Brazil, 3i Editora, 1 ed., vol 1, pp 169–193

Ruchkys UA, Machado MMM (2015) Serviços da Geodiversidade Associados às Formações Ferríferas: Pressão e Oportunidades de Conservação. In: Patrimônio Espeleológico em rochas ferruginosas: propostas para sua conservação no Quadrilátero Ferrífero, Minas Gerais. Ruchkys UA, Travassos LEP, Rasteiro MA, Faria LE (Orgs.), Campinas, SP, Sociedade Brasileira de Espeleologia (SBE), 1 Ed., pp 262–273. Available for download at http://www.cavernas.org.br/editorasbe/perferruginosas.pdf. Accessed 07 Feb 2018

Salgado AAR, Carmo FF (2015) Quadrilátero Ferrífero: a beautiful and neglected landscape between the gold and iron ore reservoirs. In: Vieira BC, Salgado AAR, dos Santos LJ (eds) Landscapes and landforms of Brazil, World geomorphological landscapes. Springer Science+Business Media, Dordrecht, pp 319–330

Salgado-Labouriau ML, Ferraz-Vicentini KR (1994) Fire in the Cerrado 32,000 years ago. Curr Res Pleistocene 11:85–87

Santos BA (1986) Recursos minerais. In: Almeida JMG Jr. (ed) Carajás: Desafio Político, Ecologia e Desenvolvimento. Brasília/CNPq; Brasiliense, São Paulo, pp 294–361

Santos JLR (2010) Território em transe: a Floresta Nacional de Carajás. In: Anais from the V Encontro Nacional da Associação Nacional de Pós-Graduação e Pesquisa em Ambiente e Sociedade (ANPPAS), Florianópolis, Brazil. CDROM, GT16-43-16-20100903195758, available for download at http://www.anppas.org.br/encontro5/cd/artigos/GT16-43-16-20100903195758.pdf. Accessed 07 Feb 2018

Schaefer CEGR, Mendonça BAF, Ribeiro AS (2008) Solos desenvolvidos sobre canga ferruginosa no Quadrilátero Ferrífero, Minas Gerais. In: Jacobi CM (ed) Simpósio Afloramentos Ferruginosos no Quadrilátero Ferrífero: Biodiversidade, Conservação e Perspectivas de Sustentabilidade. Universidade Federal de Minas Gerais (UFMG)/Instituto de Ciências Biológicas (ICB), Belo Horizonte, pp 109–121

Schaefer CE, Cândido HG, Corrêa GR, Pereira A, Nunes JA, Souza OF, Marins A, Fernandes Filho E, Ker JC (2015) Solos desenvolvidos sobre Canga Ferruginosa no Brasil: Uma Revisão Crítica e Papel Ecológico de Termiteiros. In: Geossistemas Ferruginosos do Brasil: áreas prioritárias para conservação da diversidade geológica e biológica, patrimônio cultural e serviços ambientais. Belo Horizonte, Brazil, 3i Editora, 1 ed., vol 1, pp 77–102

Silva HZ (2001) Hidrogeologia de regiões mineiras. REM: Int Eng J, Rev Escola de Minas 54 (3):199–204

Souza FCR, Carmo FF (2015) Geossistemas Ferruginosos no Brasil. In: Geossistemas Ferruginosos do Brasil: áreas prioritárias para conservação da diversidade geológica e biológica, patrimônio cultural e serviços ambientais. Belo Horizonte, Brazil, 3i Editora, 1 ed., vol 1, pp 47–76

Takahasi A (2015) Flora das Cangas de Corumbá, MS: Diversidade e Conservação. In: Geossistemas Ferruginosos do Brasil: áreas prioritárias para conservação da diversidade geológica e biológica, patrimônio cultural e serviços ambientais. Belo Horizonte, Brazil, 3i Editora, 1 ed., vol 1, pp 317–334

Tavares VC, Palmuti CFS, Gregorin R, Dornas TT (2012) Morcegos. In: Martins FD, Castilho AF, Campos J, Hatano FM, Rolin SG (Orgs.) Fauna da Floresta Nacional de Carajás: estudos sobre vertebrados terrestres. São Paulo, Nitro Editorial, pp 162–179

Tobias Junior R, Nascimento ELM, Rodrigues IMM (2015) Contexto Arqueológico e Longa Duração nas Serras do Paraopeba, Negra e do Itabira, MG. In: Geossistemas Ferruginosos do Brasil: áreas prioritárias para conservação da diversidade geológica e biológica, patrimônio cultural e serviços ambientais. Belo Horizonte, Brazil, 3i Editora, 1 ed., vol 1, pp 429–463

Treece D (1987) Bound in misery and iron: the impact of the Grande Carajás Programme on the Indians of Brazil. Report from Survival International with an environmental assessment by Friends of the Earth. Colchester M, Luling V (eds). RAP Limited, UK, 151 pp

Trendall AF, Morris RC (1983). Iron-formation: facts and problems. In: Trendall AF, Morris RC (eds) Developments in Precambrian geology, vol 6. Elsevier Science Publishers B.V., Amsterdam, 558 pp

Vasconcelos MF, Rodrigues M (2010) Patterns of geographic distribution and conservation of the open-habitat avifauna of southeastern Brazilian mountaintops (campos rupestres and campos de altitude). Papéis Avulsos de Zoologia, São Paulo 50(1):1–29

Vasconcelos MF, Hoffmann D (2015) Avifauna das Vegetações Abertas e Semiabertas Associadas à Geossistemas Ferruginosos do Brasil: Levantamento, Conservação e Perspectivas para Futuros Estudos. In: Geossistemas Ferruginosos do Brasil: áreas prioritárias para conservação da diversidade geológica e biológica, patrimônio cultural e serviços ambientais. Belo Horizonte, Brazil, 3i Editora, 1 ed., vol 1, pp 259–287

Zhao J, Huggins FE, Zhen F, Huffman GP (1994) Ferrihydrite: surface structure and its effects on phase transformation. Clays Clay Miner 42(6):737–746

Chapter 6
Gigantic Cities and the New Challenge of an Old Science: Geoethics, Geoeducation, and Geoknowledge in Porto Alegre, Brazil

Rualdo Menegat and Rodrigo Cybis Fontana

> *Man's advent has not been a mere solitary fact, nor have the alterations which he has effected been confined solely to the relations that subsist between himself and nature. He has set in motion a series of changes which have reacted on each other in countless circles, both through the organic and inorganic world. Nor are they confined to the past; they still go on; and, as years roll away, they must produce new modifications and reactions, the stream of change ever widening, carrying with it man himself, from whom it took rise, and who is yet in no small degree involved in the very revolutions which he originates.*
> Sir Archibald Geikie (1901, p. 425.)

Abstract In the present time, we are facing new challenges that require us to review the role of Geology to the material, cognitive and ethic progress of humankind. The vertiginous population growth, urban sprawl, increasing demand for solid, liquid, and gaseous geosphere, the collapse of the ecosystems that support life and the climate change replace Geology as one of the most important contemporary sciences. We need to review procedures, analytical scales and contribute to civilization process. Essentially, we need Geology help develop a deep Geoethic on the use of land resources, care with the geosphere, and mankind sustainability. The Geoethic emerges as a new understanding that cleaves across all geological activities, whether of scientific practice, or the practice of exploitation of natural

R. Menegat (✉) · R. C. Fontana
Instituto de Geociências, Universidade Federal do Rio Grande do Sul,
Porto Alegre, Brazil
e-mail: rualdo.menegat@ufrgs.br

R. Menegat
International Association for Geoethics (IAGETH), Brazilian Chapter,
Porto Alegre, Brazil

R. Menegat
Foro Latino-Americano de Ciencias Ambientales (FLACAM),
Cátedra UNESCO/Unitwin para el Desarrollo Sustentable,
Ciudad de La Plata, Argentina

R. D. Acevedo and J. M. Frías (eds.), *Geoethics In Latin America*,
The Latin American Studies Book Series,
https://doi.org/10.1007/978-3-319-75373-7_6

resources, and or also the knowledge of how geodynamics affects cities and human activities and vice versa. At the root of contemporary environmental and urban problems, whether planetary or local is how the city is perceived by its inhabitants, managers, and intelligentsia. The ascent of urban realm in the next 30 years is expected to almost duplicate the nowadays 3.6 billion inhabitants. The central challenge for the geology and Geoethics is to consider the many problems due to the complex relations between the large urban needs and geolandscape transformations. The main goal of this chapter is to describe the ascent of cities' scales—from city and megacity to megalopolis and ecumenopolis—in order to describe connections between the physical world-city and the superficial components of the Earth systems. The techno-urbansphere is defined as the urban physical totality, which includes the man-made system and the lithosphere, hydrosphere, atmosphere, and biosphere's portions transformed by it. Due to geologic scale of the techno-urbansphere, it is not possible to observe it by a citizen without technical and Earth science concepts and instruments. To offer to citizens and decisions makers accurate instruments to understand the nowadays urban geocomplexities, thematic surveys of the urban physical totality are very important. This possibility is illustrated by the Environmental Atlas of Porto Alegre case, which triggered new looks in urban environmental management, geoethics and geoeducation in Porto Alegre city, Southern Brazil.

Keywords Urban sprawl · Geohazard · Urban sustainability · Geoethic culture
Urban geological scale · Urban environmental management · Geosphere
Techno-urbansphere

6.1 Introduction

Through time, Geology has been recognized for contributing to the material success of humankind. Though not always with the name "Geology," coined by Ulisse Aldrovandi in 1603 (Vai 2004), this science and how to know the mineral kingdom, also referred in the past as oryctology, geomancy, geognosy, among others, is one of the oldest knowledge of humankind. The first human groups, like Homo habilis (2.5 Ma), knew collect rocks suitable for the manufacture of instruments. Minerals like obsidian and flint provided support for the development of advanced tools, as the famous flint dagger of Çatalhöyük (Hodder 2013), in Anatolia (7500 BC). Civilizations have been known due to the mineral and rock that support of its buildings and artifacts: clay for Mesopotamia and Yellow River, granite and limestone for ancient Egypt and Incas in the Central Andes, marble for ancient Greek and Rome, or sandstone for Tiwanaku in the Titicaca Lake region. In fact, the periods of human history are delimited according to the use of earthly goods, as Iron Age, Bronze Age, Atomic Age, Oil or Silicon Age. The very notion of historical memory of human cultures also depends on the materials: we comment much civilization who used rocks and whose records have not been destroyed by time—like the pyramids of Egypt—than the Amazonian cultures, whose settlements and artifacts wood have deteriorated.

The sciences and methods that enabled select rocks and minerals of the Earth's surface, not only provide the material basis to support civilizations, but also served as ways of thinking about the world. This contribution of Geology, for the spiritual progress of humankind (Menegat 2008a), is less known, even among geoscientists. Usually, the literature refers more about the cultural impact of the astronomical and biological theories or even a particular technology, without considering that the discovery of Earth's dynamics is one of the most mysterious and amazing among of all human adventures (Menegat 2008b). The progress of geological science led to surpass the mineral, vegetable, and animal fixed kingdoms of Middle Age in terms of a very old dynamic Earth system. We can now understand the evolution of life, their extinctions, deep time and inaccessible Earth's depths interior, whatever the scale. Although much unknown to the general public, theories that Geology offers are so impressive that led Darwin, one of greatest geologists (Geikie 2009 [1909]; Herbert 2005), claiming that it creates the same great visions of the Earth that astronomy in relation to the universe.

However, at the present time, we are facing new challenges that require us to review the role of Geology to the material and spiritual progress of humankind. The vertiginous population growth, urban sprawl, increasing demand for solid, liquid, and gaseous geosphere, the collapse of the ecosystems that support life and the climate change replace Geology as one of the most important contemporary sciences. We need to review procedures, analytical scales and contribute to civilization process. Essentially, we need Geology help develop a deep Geoethic on the use of land resources, care with the geosphere, and mankind sustainability. The Geoethic emerges as a new vision that cleaves across all geological activities, whether of scientific practice, or the practice of exploitation of natural resources, and or also the understanding of how geodynamics affects cities (Coch 2004; King et al. 2007) and human artifacts and vice versa. The Geoethic puts the need to enter the Geology and Earth themes in culture. That is to say that the understanding of the place in which we live in should be part of the basic curricula of the citizens anywhere in the world, especially in large cities.

Geology arises as a science that helps to evaluate all aspects of the relationship between humankind activities and terrestrial systems, including cognitive and ethical dimensions. In this chapter, we discuss how Geology is fundamental to understand the new equation of natural history: to know the place of cities in Earth system and the place of citizens in cities, so that all practices could be reviewed in terms of Geoethics. At the root of contemporary environmental and urban problems, whether planetary or local, is how the city is perceived by its inhabitants, managers, and intelligentsia. Although the city is the most complex human artifact (Mumford 1961; Girardet 1992; Batty 2013), the urban question is usually addressed only to planners and from Geddes (1915), Chicago School (Park et al. 1925) Mumford (1938), Toynbee (1967), and Doxiadis (1968), also for sociologists, historians, and geographers (e.g., Gottdiener and Mutchison 2006; Welter 2003; Theodorson 1982).

By the 1930s decade, the city as a geographical and geological fact won contours through the books of Lewis Mumford [1895–1990] as "The culture of the cities." In this book, Mumford (1938, p. 316) stated that the city is an earthly form and expression of regional individuality. Even when it is very large, "its shape is conditioned by the topography and the nature of the land." One of the major scientific contributions of Mumford was to address the issue of the city to all human subjects and activities, due to the complexity of the urban artifact and its relationship with the landscape, technology, society, and human nature (Mumford 1944, 1961, 1964).

In later decades, the issue of cities became part of the concern of many other scientific areas such as Human Ecology, Landscape Ecology, Anthropology, and, most recently, Geology. In all of them, we will find contributions of the Scottish biologist Patrick Geddes [1854–1932] and the American Lewis Mumford. On the importance of Geology in establishing the architectural and individual characteristics of the city, Mumford has shown us that

> At no moment in existence of the city is the divorce between the man-made environment and the Earth complete. The red sandstone of Strasbourg, the yellow clay of London, the red brick of Bremen, the gray limestone of post-medieval Paris, the brown sandstone of the old Frankfort-am-Main-the very bricks and stones symbolize the underlying partnership between man and the nature, which is accepted and furthered, even while it is transformed, in the structure of the city. (Mumford 1938, p. 318)

In the inextricable link between the city and the geological environment, Mumford went further to analyze the influence of geology in shaping the consciousness of urban dwellers and the internal aspects of the city, such as the distribution of the buildings, as we can see below:

> The immediate geological foundations remain an important attribute of urban individuality; they seep into the consciousness of individuals in all sorts of indirect ways. The very infant at play, digging in the earth of his dooryard, is conscious of the ubiquitous sand if he lives in Rotterdam, of the oily shale if in Pittsburgh, or of the tantalizing gleam of mica in the schist of Manhattan. And as cities develop, external conditions become internal influences [...]. (Mumford 1938, pp. 318–19)

In fact, the interest for urban realm by geologists was advanced with the Urban Geology (GSC 2008), the new discipline that emerged in association with Engineering Geology from the 1960s, when urban gigantism already was outlined. One of the pioneering works was written in 1964 by the geologist McGill (1964) with the title Growing importance of urban geology, and published in the Bulletin of the United States Geological Survey. Four years later, the first engineering geology session took place under an International Geological Congress, which was entitled "Engineering Geology in regional planning" (Legget 1973). From this, the work on urban geology has become increasingly frequent. In 1973, the English teacher Robert Ferguson Legget [1904–1994], then a member of the National Research Council of Canada, published the classic manual Cities and Geology (Legget 1973), paving the way in defining the scope of the Urban Geology.

This chapter briefly points out about the elements that make the city a local and global geological fact. In addition, we discuss how Geology contributes not only from a technical point of view—providing raw material for construction, solutions for urban planning and various environmental problems—but also how it contributes to the scientific understanding of the giant urban world. The urban realm will be presented in terms of the traditional population and geographic scales. But we will present the geological scale: the global city as a huge physical urban system which includes, besides the buildings, the portions of the planetary spheres where cities are embedded and maintain interfaces with it. This global urban physical system (or the urban realm) is introduced as a thin terrestrial layer named as techno-urbansphere. From these premises, it will outline the importance of geoinformation systems for planning, management, and geoeducation in big cities as function of geoethic concerns. Further, we will describe the experience developed through the Environmental Atlas of Porto Alegre, South Brazil (Menegat et al. 2000, 2006).

6.2 Terra Urbis: Growth and Scales of the Cities in the Urbanizing World

The entire urban system consists of several physical components, natural and constructed, social and cultural, that all interact dynamically in the landscape over time. Usually, this totality is buried by many misreading and seen only in its simplest and obvious aspects. Cities are looked only whether by the streets and avenues tapestry, which appears in our daily lives as addresses, or by building standards that make up the mosaic of neighborhoods. Both aspects are represented by only two dimensions on maps or city plans. In terms of its physical components, the city is still referred to by the type and aesthetics of their buildings and quality of urban mobility, dominated by intensive traffic. Much of the urban master plans do not go beyond the normalization of these parameters.

Cities are also scaled by the amount of inhabitants. Throughout the twentieth century, cities have sought to be synonymous with physical and population greatness. The goal of the small towns was to be medium, and the goal of these, lay in becoming large (see Table 6.1). In turn, major cities planning to become, as soon as possible, in metropolis, in turn, these disputed for the title of "the world's largest city." In the early nineteenth century, driven by the industrial revolution, colonial wealth, and energy obtained from coal, London, became the first city to reach the figure of one million inhabitants (Girardet 1992), which until then only the ancient Rome had reached it in the first century BC (Hibbert 1987; Fleming 1991). In turn, New York was the first city to reach ten million inhabitants in the period between the world wars. Besides the size, expressed in terms of population quantity, cities also compete for height and gigantism of buildings, extension of roads and tunnels, among others. The evolution of urban growth of these indices is usually referred to as progress, a notion that endures today.

Table 6.1 The top 30 largest metropolis in the world from 1950 to 2014, prioritized by the size of its population (in millions), including metropolitan areas

	Metropolis	1950	Metropolis	1970	Metropolis	2000	Metropolis	2014
1	New York-Newark, USA	12.3	Tokyo, Japan	23.3	Tokyo, Japan	34.4	Tokyo, Japan	38.0
2	Tokyo. Japan	11.2	New York-Newark, USA	16.2	Osaka, Japan	18.6	Delhi, India	25.7
3	London, England	8.3	Osaka, Japan	15.2	Mexico City, Mexico	18.4	Shanghai, China	23.7
4	Osaka, Japan	7.0	Mexico City, Mexico	8.8	New York, USA	17.8	São Paulo, Brazil	21.0
5	Paris, France	6.2	Calcutta, India	6.9	São Paulo, Brazil	17.0	Bombay, India	21.0
6	Moscow, Russia	5.3	Rio de Janeiro, Brazil	6.8	Bombay, India	16.3	Mexico City, Mexico	20.9
7	Buenos Aires, Argentina	5.0	São Paulo, Brazil.	7.6	Delhi, India	15.7	Beijing, China	20.3
8	Chicago, USA	4.9	Bombay, India…	5.8	Shanghai, China	13.9	Osaka, Japan	20.2
9	Calcutta, India	4.5	London, England	7.5	Cairo, Egypt	13.6	Cairo, Egypt	18.7
10	Shanghai, China	4.3	Nagoya, Japan	6.6	Calcutta, India	13.0	New York, USA	18.6
11	Los Angeles, USA	4.0	Saint Petersburg, Russia	3.9	Buenos Aires, Argentina	12.4	Dhaka, Bangladesh	17.6
12	Mexico City, Mexico	3.3	Detroit, USA	3.9	Los Angeles, USA	11.7	Karachi, Pakistan	16.6
13	Berlin, Germany	3.3	Jakarta, Indonesia	3.9	Rio de Janeiro, Brazil	11.3	Buenos Aires, Argentina	15.1
14	Philadelphia, USA	3.1	Los Angeles, USA	8.3	Dhaka, Bangladesh	10.2	Calcutta, India	14.8
15	Rio de Janeiro, Brazil	3.0	Paris, France	8.2	Beijing, China	10.1	Istanbul, Turkey	14.1
16	Saint Petersburg, Russia	2.9	Cairo, Egypt	5.5	Karachi, Pakistan	10.0	Chongqing, China	13.3
17	Bombay, India	2.8	Buenos Aires, Argentina	8.1	Moscow, Russia	10.0	Lagos, Nigeria	13.1
18	Detroit, USA	2.7	Moscow, Russia	7.1	Manila, Philippines	9.9	Manila, Philippines	12.9
19	Boston, USA	2.5	Chicago, USA	7.1	Seoul, S. Korea	9.8	Rio de Janeiro, Brazil	12.9
20	Cairo, Egypt	2.4	Seoul, S. Korea	5.3	Paris, France	9.7	Guangzhou, China	12.4
21	Tianjin, China	2.4	Beijing, China	4.4	Istanbul, Turkey	8.7	Los Angeles, USA	12.3
22	Manchester, England	2.4	Manila, Philippines	3.5	Nagoya, Japan	8.7	Moscow, Russia	12.1
23	São Paulo, Brazil	2.3	Delhi, India	3.5	London, England	8.6	Kinshasa, Congo	11.5

(continued)

Table 6.1 (continued)

	Metropolis	1950	Metropolis	1970	Metropolis	2000	Metropolis	2014
24	Nagoya, Japan	2.2	Madrid, Spain	3.5	Jakarta, Indonesia	8.3	Tianjin, China	11.2
25	Birmingham, England	2.2	Philadelphia, USA	4.3	Chicago, USA	8.3	Paris, France	10.8
26	Shenyang, China	2.1	Barcelona, Spain	3.4	Chongqing, China	7.8	Shenzhen, China	10.7
27	Rome, Italy	1.9	Hong Kong, China	3.4	Guangzhou, China	7.3	Jakarta, Indonesia	10.3
28	Turin, Italy	1.9	Shanghai, China	6.0	Lima, Peru	7.2	London, England	10.3
29	Milan, Italy	1.9	Tianjin, China	3.3	Lagos, Nigeria	7.2	Bangalore, India	10.0
30	Naples, Italy	1.9	Kitakyushu, Japan	4.0	Tehran, Iran	7.1	Lima, Peru	9.8

Data Source United Nations (1985), (2015)

6.2.1 The Metropolis Without Limits: The Era of the Giant Cities

This logic of urban growth, which was inflation during the nineteenth century in more developed regions, and XX, in less developed regions, has led to increased urban world population (United Nations 2007, 2009; Angel et al. 2012; Angel 2012; Knox and Florida 2014) and the disproportionate gigantism and unprecedented towns (Bell and Tyrwhitt 1972a, Dogan and Kasarda 1988a, b; Mumford 1961; Hardoy et al. 2001; Montgomery et al. 2004). In 1950, there were ten cities with more than five million inhabitants, of which the first three—New York, London, and Shanghai—have housed more than ten million. As can be accompanied in Tables 6.1 and 6.2, in 1985, the number of megacities jumped to 34. In 2014, the number of megacities was already in the house of 70 (Brinkhoff 2015; United Nations 2015), of which 32 with more than 10, 13 with more than 20, and 3 over 30, and 1 over 40 million. This trend will continue to increase even more in the coming decades. Projections indicate that by 2030, there should be about 104 megacities (United Nations 1985, 1986; Dogan and Kasarda 1988c), which concentrate a population of 1164 million (see Table 6.2).

For the first time, in 2007 the urban population of the planet surpassed the rural population (United Nations 2015). Between 2014 and 2050, it projected a growth of the world population of 2.3 billion people, from 7.2 to 9.5. In the same period, the urban population will increase 2.5 billion inhabitants, from 3.8 in 2014 to 6.3 billion in 2050 (United Nations 2015). Thus, the next three and half decades, the global urban population will nearly increase 66% (Fig. 6.1), becoming almost 70% of the total population (Table 6.4). To absorb this increase in urban population, it is estimated that the number of large cities and megacities will be close to no less than a hundred (Brinkhoff 2015; United Nations 2015). Much of this growth will occur in cities of less developed regions (see Fig. 6.2 and Table 6.3). It is estimated that in Asia the urban population is expected to increase 1.5, while in Africa, 0.9, and in Latin America and Caribbean, 0.2 billion people during the same period (see Table 6.4).

Table 6.2 Total population and number of urban agglomerations per size category, from 1970 to 2015 and projections to 2030

Year	Population by size category (in million)				Total urban population (in million)	Number of urban agglomeration by category			
	1–5	5–10	>10	Total		1–5	5–10	>10	Total
1950	128.8	32.1	23.6	184.5	746	70	5	2	77
1970	244.8	106.9	54.7	406.4	1350	126	15	3	144
1990	459.5	157.0	152.6	769.1	2285	239	21	10	270
2015	847.2	306.8	471.3	1625.3	3880	428	44	29	501
2030	1127.8	433.9	729.9	2291.6	5058	558	63	41	662

Source United Nations (2015)

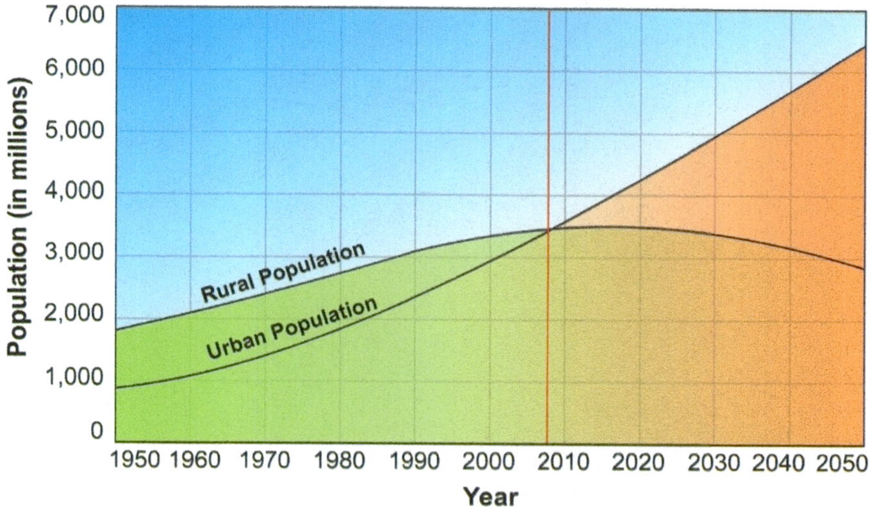

Fig. 6.1 Urban and rural population in the world, 1950–2050 *Credit* United Nations (2015)

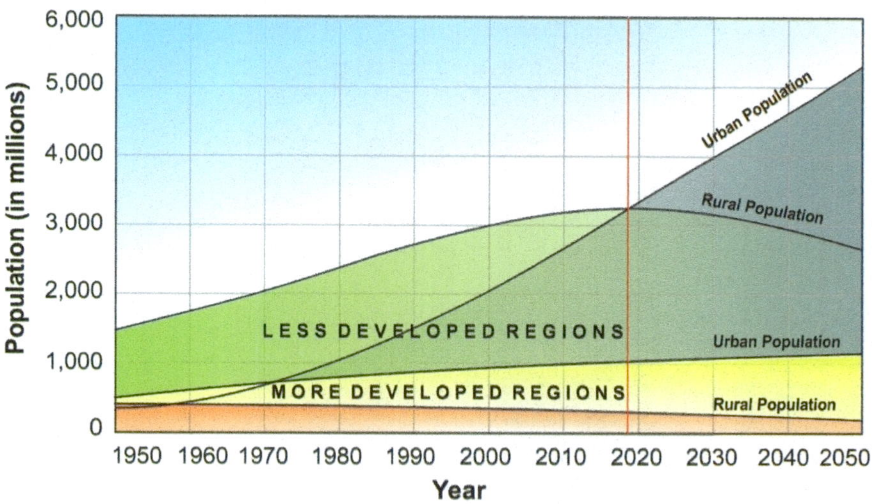

Fig. 6.2 Urban and rural populations by development region, from 1950–2050 *Credit* United Nations (2015)

This excessive growth in such a short time has put into question the very definition of "megacity" whose term is also used interchangeably with "supercity," "giant city," "conurbation" (Bell and Tyrwhitt 1972a), and "megalopolis" (Gilbert 1996). The criterion of Dogan and Kasarda (1988a) includes in this category urban

Table 6.3 Total world population, urban and rural development by region in 1950, 1970, 1990, 2014, and projections for 2030 and 2050

Region area	Population (billions)						Average annual rate of change (%)				
	1950	1970	1990	2014	2030	2050	1950/1970	1970/1990	1990/2014	2014/2030	2030/2050
Total Population World	2.53	3.69	5.32	7.24	8.42	9.55	1.90	1.83	1.29	0.94	0.63
More developed regions	0.81	1.01	1.15	1.26	1.29	1.30	1.08	0.65	0.37	0.18	0.04
Less developed regions	1.71	2.68	4.17	5.99	7.13	8.25	2.24	2.21	1.50	1.09	0.73
Urban Population World	0.75	1.35	2.29	3.88	5.06	6.34	2.96	2.63	2.21	1.66	1.13
More developed regions	0.44	0.67	0.83	0.98	1.05	1.11	2.08	1.06	0.69	0.45	0.27
Less developed regions	0.30	0.68	1.45	2.90	4.00	5.23	4.04	3.82	2.88	2.02	1.33
Rural Population World	1.78	2.34	3.04	3.36	3.37	3.21	1.37	1.30	0.43	0.01	−0.23
More developed regions	0.37	0.34	0.32	0.28	0.24	0.19	−0.47	−0.28	−0.58	−0.87	−1.18
Less developed regions	1.41	2.01	2.72	3.09	3.13	3.02	1.76	1.52	0.53	0.08	−0.17

Data Source United Nations (2015)

Table 6.4 Percentage of urban population by major area in 1950, 1990, 2014 and projections for 2030 and 2050

Major area	Total population (millions)					Urban population (millions)					Percentage urban				
	1950	1990	2014	2030	2050	1950	1990	2014	2030	2050	1950	1990	2014	2030	2050
World	2527	5320	7243	8424	9551	746	2285	3880	5058	6339	29.6	42.9	53.6	60.0	66.4
Africa	229	630	1138	1634	2393	32	197	455	770	1 339	13.9	31.3	40.0	47.1	58.5
Asia	1396	3213	4342	4887	5164	245	1 036	2 064	2 752	3 313	17.5	32.2	47.5	56.3	64.1
Europe	549	723	743	736	709	283	506	545	567	581	51.5	70.0	73.3	77.0	82.0
Lat. America Carib	168	445	623	717	782	69	314	496	595	674	41.1	70.6	79.6	83.0	86.2
Northern America	172	282	358	403	446	110	213	292	340	390	63.9	75.5	81.6	84,4	87.4
Oceania	13	27	39	47	57	8	19	27	34	42	64.5	70.4	69.2	72.3	73.7

Data Source United Nations (2015)

agglomerations with more than four million inhabitants. However, other studies reserve the term "supercity" for clusters with more than five million (Lowder 1987) or, "megacity" (United Nations 2015) when any continuous urban area has more than ten million inhabitants or, still, "megalopolis" (Mayhew and Penny 1992). For this term "megalopolis," we discuss below a category of urban agglomeration constituted by a chain of megacities.

6.2.2 *Chains of Giant Cities: How Big Can Be Urban Agglomeration?*

Megacities, in turn, tend to cluster, forming huge urban chains, corridors or aprons, which are called megalopolis (Geddes 1915; Mumford 1938; Gottman 1961; Doxiadis 1967; Papaioannou 1964; Bell and Tyrwhitt 1972a; Gottman and Harper 1990; Barker and Sutcliffe 1993; Harrison and Hoyler 2015a). Many authors as Knox and Florida (2014) name megalopolis as megaregion, which can sprawl with three forms: a long corridor or chain along a main road; as an apron, or as a circular patch. They explain that the megaregions are not a simple agglomeration, but an urban web with strong economic activity.

The four largest nowadays never-ending cities impress by the magnitude of their geographical extension and population concentration: each one have almost the impressive population of seventy million inhabitants. The first megalopolis recognized is located in the northeastern coastal region of USA. It encompasses the set of cities that have over a corridor of about 820 km in length, 100 km of medium width, and 134,680 km^2 in area, running from Boston to Washington DC (the Boswash, see Lakshmanan et al. 2015), including also the cities and megacities like Providence, Hartford, New York, Philadelphia, and Baltimore (National Geographic 1994; Short 2007). This extensive and populous urban chain contains 52.3 million people, which occupies the lowlands between the sheathed coastal waterfront (most along the bay headline) and the elevations of the Northern Appalachians can be well visualized in Fig. 6.3. This urban chain was pioneered diagnosed in 1961 by Gottman (1961). In describing this urban phenomenon, then with population and extend much smaller than the current, put the name of "megalopolis." In the 1980s, the constellation of cities in this region was still a doubt, as Professor David Leveson (1980, p. 5, emphasis added) wrote: "The urbanized areas along the US northeast coast *may eventually* coalesce to form a megalopolis," but nowadays is a fact (Short 2007; Mitchell and Leen 2001).

The European urban megalopolis can be identified by a long apron from Rhine-Ruhr metropolitan region (Bonn–Köln–Düsseldorf–Duisburg–Dortmund) (Grier et al. 2002; Lüttich 2010), undergoes to the so-called Randstad (or the "border city" that encompass Amsterdam, Rotterdam, The Hague and Utrecht), and Brussels–Ghent–Antwerp–Leuven (named as "Flemish Diamond" sensu Vanhaverbeke 1998), followed by Lille–Kortrijk–Tournai (known as "Eurometropolis'). In Great Britain, go

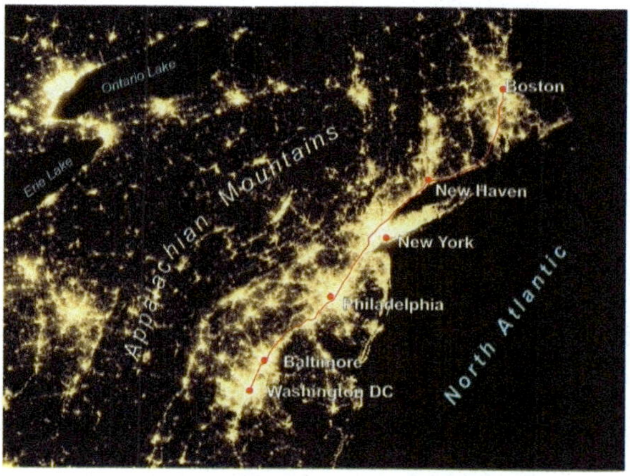

Fig. 6.3 Megalopolis of Boston–New York–Philadelphia–Washington stands on the map of the urban sprawl of northeastern coastal region of USA. The map was drawn from the satellite images composition captured at night *Credit* NASA-NOAA Satellite 2012

on to "London commuter belt," reach at north Birmingham–Nottingham–Leicester–Derby in the Midlands, and follow to Leeds–Manchester–Liverpool (the so-called M62 motorway corridor) (Bell and Tyrwhitt 1972a; Lüttich 2010). The Britain and Continental counterparts are connected via the Eurotunnel under the 50.5 km in length English Channel. This urban apron is based along the outer end Rhine valley, the peat and clay lowlands of Rhine–Meuse–Scheldt tidal delta (the "deltametropolis" sensu Frieling 2000), and the eastern dune coast of the Strait of Dover in the North Sea continental border. The Britain counterpart of the megalopolis occupies the inner end of Thames Estuary running to north until Manchester–Liverpool axis which is settled along the banks of Irwell and Mersey rivers. The main axis of the entire huge apron is about 700 km in length, and circa of 100.81 thousand km^2 in area, and home 71.3 million inhabitants.

The Japanese megalopolis is formed by the urban agglomerate chain from Saitama, Tokyo–Yokohama to the north (Okata and Murayama 2011) extending to the south to Osaka–Kobe in an extension of about 600 km (Bell and Tyrwhitt 1972b; Nagashima 1972; Ito 1980). This chain, also known as Tōkaidō ("East Coast Road") or Taiheiyō belt ("Pacific Belt"), develops in the narrow strip of lowland bounded by mountain slopes in the west and the jagged coastline of bays to the east. The narrow corridor encompasses the Greater Tokyo Area (Yokohama, Kawasaki, Sagamihara, Chiba, and Saitama), Greater Osaka urban area (Osaka, Kyoto, Kobe, Sakai, and Higashiosaka), and Chūkyō Metropolitan Area (Nagoya, Aichi, Gifu, and Mie), with 66.33 million inhabitants in a region about 140,000 km^2 in area.

Finally, the fourth greatest megalopolis, called as Pearl River Delta Megalopolis, encompasses the main cities of Guangzhou (or Canton)–Dongguan–Foshan–Shenzhen–Hong Kong sprawled on the deltaic lowlands of the Pearl River in southern

Fig. 6.4 Composite satellite image of ecumenopolis resulting from satellite images composition obtained at night between 1994 and 1995. This Mundus city homes to about 3.8 billion inhabitants, more than half the world's population
Credit NASA-NOAA Satellite 2012

coast of China (Kraas et al. 2014; Zhang 2015b). This huge urban constellation home circa of 54.02 million people (OECD 2015) along a complex deltaic apron 220 km in length from Guangzhou to Hong Kong constituted by many flat islands.

Nowadays, there are several megalopolises with more than twenty million people (see Fig. 6.4). Besides those described above, include: the Great Lakes megalopolis (Chicago–Detroit–Cleveland–Pittsburgh) (Glass 2015; Thün et al. 2015), the Shanghai–Nanking axis or Yangtze River Delta Megaregion (Yang 2009), Beijing–Tianjin axis, and surrounding areas (Bell and Tyrwhitt 1972a; World Bank 2015; OECD 2015; Harrison and Hoyler 2015a; Yang 2009). In the coming decades, there will be the tendency to arise, many other megalopolis (Bell and Tyrwhitt 1972a; Knox and Florida 2014; Kraas et al. 2014; Harrison and Hoyler 2015b). On the west coast of the USA, for example, the corridor running from Los Angeles to San Francisco presents this strong vocation (Zellner and Ruby 2000). In South America, there is the cord that extends from the Rio de Janeiro to São Paulo-Campinas and may involve Belo Horizonte.

6.2.3 The Nature Surrounded by the City: The Ecumenopolis or the Mundus City

The megalopolises or huge urban regions are better visualized in the world map composed by numerous satellite images obtained at night and selected cloudless (see Fig. 6.4). This image is one of the most significant present-day scientific products. It can be compared in importance with the first photograph of the whole Earth obtained by the astronauts of Apollo 17 or with the first world maps that represented a new cartography after the discovery of the Americas, in the Renaissance. On the world map of the urban lights on Earth, megalopolises are shown connected as a network of cities covering the continents. The regions with absence of light identify places inhabited due to very inhospitable environment. This impressive map of the Mundus city shows the geological scale of the urban humankind. This means that great urban regions embrace the entire geological environments, as the lowlands along a volcanic arc ridge or geological faults, fluvial valleys flanked by steeper slopes, floodable deltas plains, embayed coastal plains, floodable lowlands of coalescent distal fans. Much of these regions are located along boundaries of tectonic plates or along volcanic arcs, with very intense geological activity as volcanism, earthquake, tsunamis, huge slumps, sea wave storms, and floods.

The Mundus city or the planetary urbanization was first proposed in 1961 by the influential Greek city planner Constantinos Doxiadis [1913–1975] as a futuristic projection (see Fig. 6.5). Through a visionary way, he reasoned that there would be a cap on population growth, given the finite resources of the planet, stipulated between 20 and 50 billion people. This limit should, according to his predictions, be achieved at the end of the twenty-first century. Doxiadis (1967 p. 352) proposes, furthermore, that "as the city population reached its maximum, and also its maximum physical dimension, it would be in equilibrium."

With the generous name of *ecumenopolis*, he described the future Mundus city:

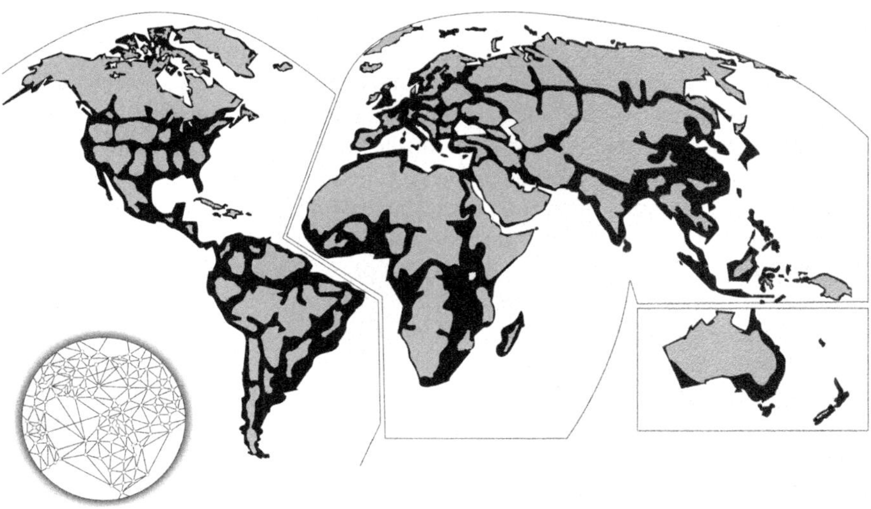

Fig. 6.5 Dark area identifies the projection of ecumenopolis for the end of the XXI century, as previewed by Constantinos Doxiadis in 1961. The image of the modern Mundus city (see Fig. 6.4), with 3.8 billion people, would correspond to an intermediate stage of this projection, scheduled for 20 billion. The globe left shows the structured network of the ecumenopolis where cities occupy different hierarchies (see Fig. 6.6) After Doxiadis (1967)

> The city of the future will take the form of a network of global dimensions. The greater hierarchy centers of the web will primarily be located where today meet the highest population concentrations, i.e., in the largest plains where are located the best water resources. The connections between them follow the natural lines of communications as well as the submarine and underground tunnels and the corresponding air corridors (Doxiadis 1967, p. 352).

Later, in 1974, he published a book in which he claimed to be the *ecumenopolis* an inevitable fact (Doxiadis and Papaioannou 1974). He imagined it as a hierarchically structured network from major urban centers to smaller Fig. 6.5). While the larger centers correspond to the nowadays immense metropolitan areas—the megacities—with a population from 5 to 10 million inhabitants, the higher-order urban region would reach a population of hundreds of millions—the megalopolis—(see Fig. 6.6).

6.3 The City as Geological Global Fact: e Dawn of the Techno-Urbansphere

Although Doxiadis (1967; Doxiadis and Papaioannou 1974) has advanced in building an integrated view of urban system in both regional scale and in its connections with natural systems, he described the ecumenopolis mainly in his two-dimensional aspect, i.e., as growing urban area due to increased population. However, this thin platform of widespread artificial rocks over the globe in the form of a network can be described, from a geological point of view, in three and, including time, in four dimensions. In addition, it can be considered in terms of the surface and undersurface dynamics interactions with the major components of the of the Earth system, namely the lithosphere, atmosphere, hydrosphere, and biosphere.

That is, in the context of the Earth system, the Mundus city or ecumenopolis can be understood as a global layer and properly called as *techno-urbansphere*. In this three-dimensional layer, with a complex geometry of interfaces, are included: (a) the physical urban artifact, or built system, consisting predominantly of concrete (crushed stone, sand, and cement), bricks, iron, glass, copper, and asphalt; (b) dynamic urban processes driven by complex techno-industrial, chemical, electronic, magnetic, and atomic systems; (c) the portions of the modified planetary spheres, assimilated, contaminated, metabolized, and appropriate by the city over time, in local and regional scales, commonly called domestication of nature (see Figs. 6.7 and 6.8).

In fact, the city is not only "supported" by the ground in the same way that a book rests on a table: It embeds in the geological underground, where establishes an interface edge with solid, liquid, and gaseous complex reactions. This interface edge between the city and the lithosphere acquires some city's artifacts and properties. The tunnel, for example, no more pertains to the mountain, but to the route that connects cities. Similarly, when the city dumps its sewage in the water of rivers, lakes, and oceans, it goes to incorporate oneself in these water bodies. The metropolis needs them both to water supply its citizens and activities, and to dump

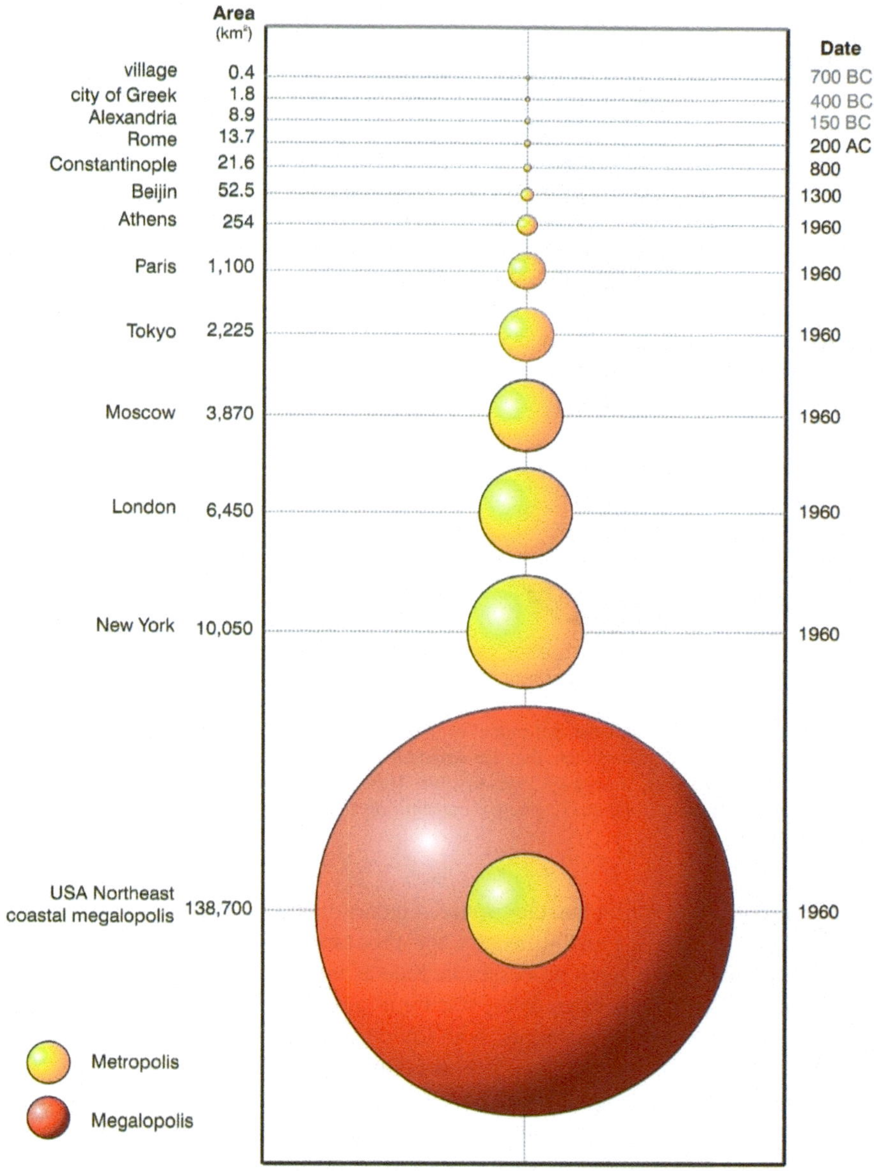

Fig. 6.6 Doxiadis' Scale for cities size, being the smallest to largest: village, town, city, megacity, megalopolis, and ecumenopolis (not shown) After Doxiadis (1967)

their high reactivity leftovers. Therefore, these contaminated water bodies are an inseparable part of the urban environment, with smells, colors, and features that are typical of the urban realm.

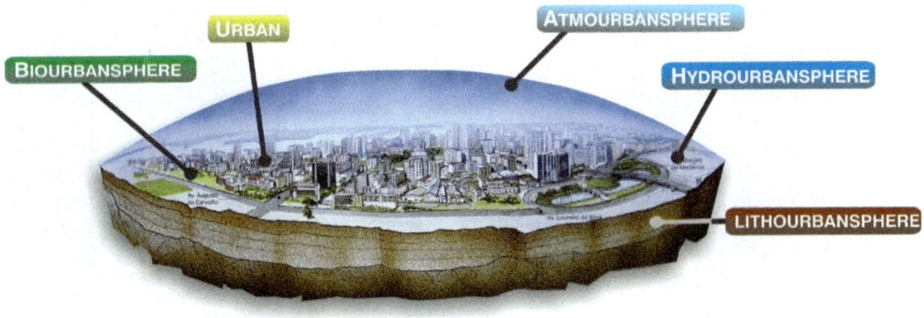

Fig. 6.7 Three-dimensional model of techno-urbansphere, identifying the portions of each planetary sphere changed and caught by it *Credit* Menegat and Fontana

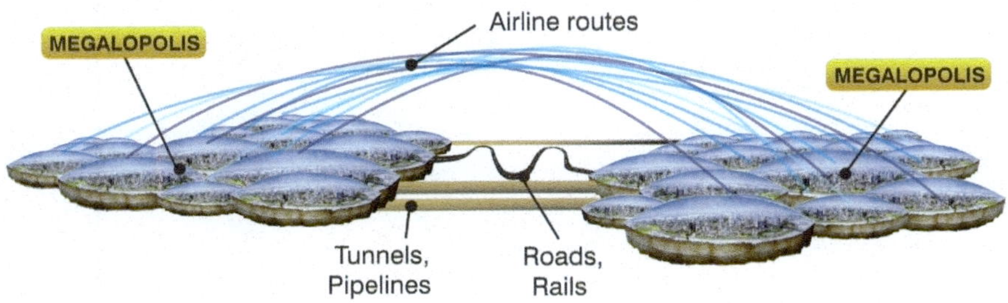

Fig. 6.8 Designed three-dimensional model of the techno-urbansphere. Megacities are connected to each other by numerous artifacts that are embedded in the planetary spheres, such as tunnels, pipelines, roads, railways, waterways, and airways. The global urban physical system includes the portions of the modified planetary spheres, constituting the techno-urbansphere *Credit* Menegat and Fontana

These portions of the planetary spheres incorporated by the Mundus city as an interface are intensely transformed by the processes that physically support and supply the inhabitants. Such processes include as the extraction of abiotic (mineral, water, air) and biotic resources as the release into the environment of many types of solid, liquid, and gaseous materials, both toxic and inert. These wastes are expelled continuously by urban metabolic processes (Girardet 1992; Adler and Tanner 2013; Karvounis 2015; Lietzke et al. 2015), which come to form significant deposits of solids and liquid and gaseous plumes. On the local scale, such waste and processes seem to be out of the urban dwellers' concerns. Many authors have defined human action in nature as a geological event in the Anthropocene geological time (Alasiewicz et al. 2012), given the magnitude of these deposits and the transformation caused by urban metabolism (e.g., Fyfe 1991; Goudie 2000; Syvitski et al. 2005).

Thus, the part of the lithosphere, including soils, modified, and incorporated by the ecumenopolis is the litho-urbansphere (Fig. 6.7). This man-made substrate can be in some places up to 6 km thick (like some oil exploration wells or in mines). The evolution of cities follows *pari passu* the development of mining and resources exploration. The higher urban growth follows the higher mineral production (Chew 2001). In the era of megacities, the lithosphere is drilled and carved into ever-deeper levels, as to extract mineral and energy resources (coal, oil and gas), as to build tunnels, urban infrastructure, and disposal material in technogenic deposits (Peloggia 1998; Carvalho 1999; Meuser 2012).

Likewise, surface and groundwater transformed by domestic, commercial, and industrial activities of the city are the hydro-urbansphere (Fig. 6.7). The North Sea and Baltic waters on the northern coast of Europe, for example, are so severely polluted (National Geographic 1992; Goudie 2000) that cannot be seen as belonging to the natural hydrosphere, but as man-made liquid edge of the northern European portion of the Mundus city.

Similarly, the modified air by domestic, industrial, and vehicular urban emissions constitutes the atmo-urbansphere (Fig. 6.7). The area affected by acid rain resulting from air pollution by US eastern cities extends to the Mississippi region (Goudie 2000). In Brazil, this area extends from the cities of Rio de Janeiro and Sao Paulo to the far west Pantanal (National Geographic 1992). The growth of atmo-urbansphere (or man-made atmosphere) has been an increasingly worrying issue due to greenhouse gas streams emitted by the activities of the Mundus city. According to Lietzke et al. (2015), urban atmosphere is circa of 100 km thick and can be divided into three sublayers: The lower, called as roughness layer, where microscale processes take place; the middle, known as inertial sublayer that refers to local scale processes; and the upper, named as outer urban boundary layer, include mesoscale of the region. Climate change is one of the symptoms of how cities appropriate themselves the atmosphere and landscape (see Pésci 2006; Bartlett and Satterthwaite 2016). In this sense, it is not the main disease, but the effect of the urban metabolism that emits greenhouse gases (see Chrysoulakis and Castro 2014).

Finally, the intense change of biota by urban needs and activities originates the bio-urbansphere (Fig. 6.7), where the chemical–mechanical intensive agriculture and biodiversity loss are some of its transformations. To be a great source to support human life (Williams 2002), the depletion of the biosphere, as evidenced by the maps of the distribution of natural forests in the USA between 1620 and 1920 (Fig. 6.9), has been the first one to get the attention of social groups, ecological and biological scientists, and some institutions. Thomas Malthus, in 1798, dealt with the possibility of an exhaustion as food supply crisis, when compared the geometric growth of the population with the mere arithmetic growth of the agricultural production. But the problem of the biosphere depletion gained wider dimensions with the pioneering works of Rachel Carson, by the impressive book *Silent Spring* (Carson 1962), Jean Dorst, who wrote in 1965 the book *Before Nature Dies* (Dorst 1965), and the MIT team headed by Donella Meadows, who wrote *The Limits to Grow* (Meadows et al. 1972). However, the broader picture of the relationship of the biosphere with human actions and other Earth systems has been offered by Gaia theory of Lovelock (1979), which established the idea that the biosphere is an interdependent of others Earth systems.

The *techno-urbansphere* is a component of the biosphere (Fig. 6.10) with great transformation and destruction capacity of Earths systems. Some authors, as Paccini

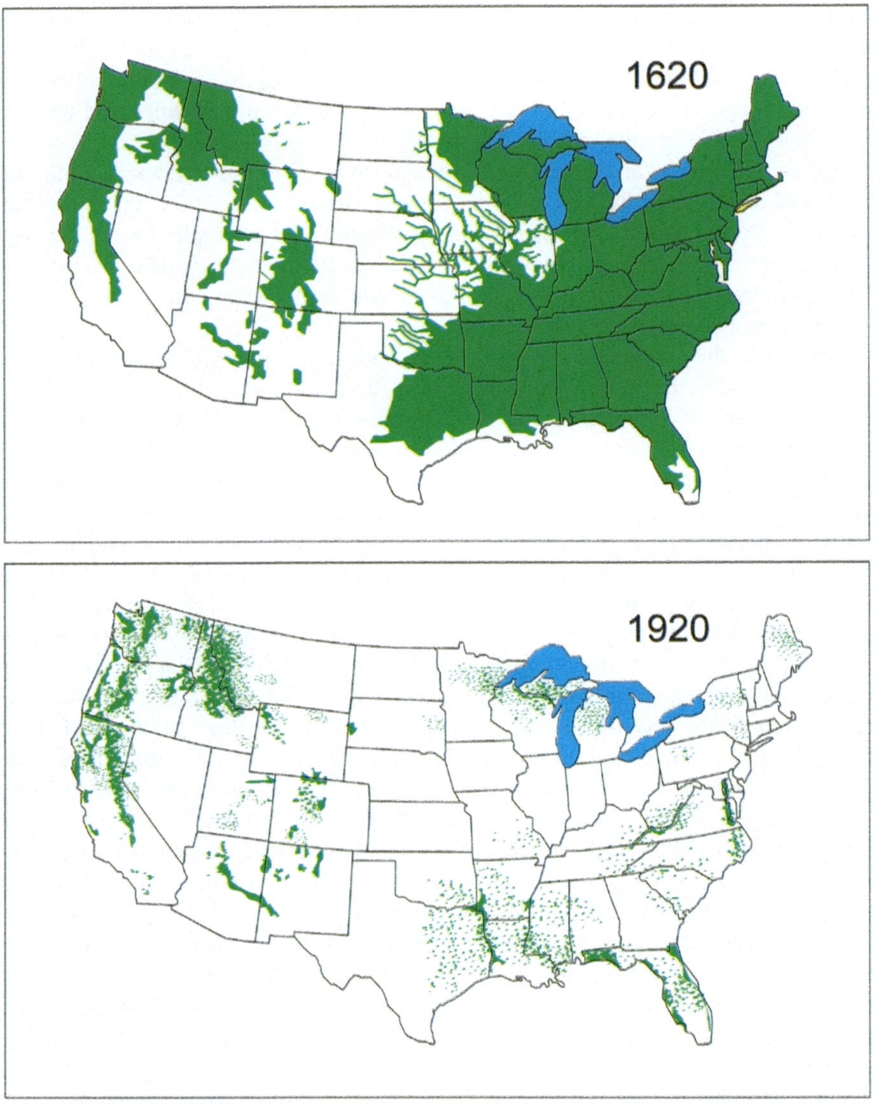

Fig. 6.9 Distribution of natural forests in the United States in 1620 and 1920. Three hundred years later, the 170 million hectares of forested in 1620, remained about 10 million
Credit Williams (1992)

and Brunner (2012), named this man-made global layer as "antroposphere." We called techno-urbansphere due to massive extra-human scale and physical complexity driven by technology, much more than the mild cities of ancient Mesopotamia, Egypt, Hindu or China, or Mesoamerica and Andes, where began the simple

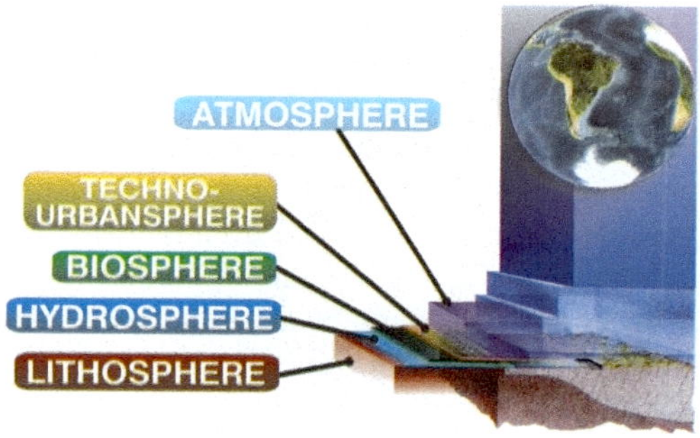

Fig. 6.10 Main components of the Earth systems, where techno-urbansphere is highlighted in yellow After Menegat et al. (2006)

urbansphere. The antroposphere—cultural, social and biological mankind—may see like the soft component of the huge physical urban layer. Figure 6.10 shows schematically the spheres of the Earth system and the place occupied in it by techno-urbansphere. In the era of gigantic cities, no longer arises over the classic question of natural history—"What's the place of humankind in nature?"—addressed since ancient Greek. The nowadays question is to investigate the place of techno-urbansphere in the Earth system in order to know the place of humankind in the city. Thus, the humankind does not have more a direct relation with Nature, as an ingeniously antroposphere, but the relation is given through a complex techno-urbansphere, defined as geological agent due to scale and long-term impact in the Earth systems.

6.3.1 *Understanding Cities as a Global Geological Fact and Introducing Geoethics: New Challenges to Live in Urban Realm*

Regardless of its range, every city and human settlement is a geological fact. The ancient cities of Mesopotamia, as the Sumerian city Ur on the Euphrates River banks, were built of sun-baked clay bricks. In addition to protecting its people against attacks from other human groups, urban construction was designed to face the furor of the Euphrates floods. Through trial and error method, the city's builders learned the first geological lessons about seasonal fluvial cycles, building materials, and flood safety techniques.

Both building materials and the choice of place to build a city demand the best of human knowledge and sound techniques (Menegat 2006). In building a city, it is not in play just the beauty of the buildings or the life protection of a single inhabitant's generation. The main question is to know how big cities could face geological processes through time if citizens do not understand about the Earth dynamic. Therefore, cities are also a human mechanism to warranty the long permanence of a culture, how they were human probes in deep natural-geological systems, in order to reach better understanding about them and so modify both culture and nature support and thereby establish more appropriate strategies for survival.

In contemporary cities where the walls of the houses are all covered with homogenized mortar and tint, citizens have lost the ability to relate the construction materials into the surrounding landscape. In addition, building materials are transported to the urban site from faraway places whose landscapes are unknown to city dwellers. Cities seem to be the result of automatic procedures of construction, whose materials come from an inexhaustible source that lies somewhere far away and independent of fortune of natural cycles.

But the materials are not the unique inexhaustible resource for the urban realm. The amount of space in which cities were established appears to have elastic properties and capacity to indefinitely house the urban growth. This illusion is partly provided by the assumption that city is a geometric board that allows its expansion in all directions without functional loss. Cities expand how if they could keep the symmetry of the board that gave rise to them and as if there were no natural barriers to stop their advance (see Fig. 6.11). As the influential philosopher Michel Serres taught, we geometrically idealize the world and our habitat and, "from that, we

Fig. 6.11 Bird's eye view of Salt Lake City (Utah, USA), 1870, made by Augustus Koch. The perfectly orthogonal urban plot infuses the never-ending space logic and the illusion that the city would not have boundaries to limit its expansion *Credit* Reps (1998, p. 97)

inhabit this space like a home, or even better, like our Earth: the meter is the Earth, and this is the profound sense of the word geometry" (Serres 2002).

6.3.2 The Geohazardous Edges of Techno-Urbansphere: When the City Reaches the Volcano Slope and Plate Tectonics Boundary

A kind of blindness emerges due to great urban expansion, and it is caused by the loss of the relationship between landscape—the total ecosystem where the city is placed—and the physical city, understood only as a geometric aesthetic. This urban blindness has pushed the city limits to distant places of the small and domestic founding core. The city was rapidly losing consciousness of its own size and of the new places that it was taking over time, known as periphery and sometimes located in very geological hazardous areas. This periphery is not only a different form of the traditional city core, but it is the possible form that human groups had developed for live in geologically dynamical areas of the edge of techno-urbansphere. It is also a loss of urban form and content as had been established in the West tradition (Menegat 2003).

In fact, Angel et al. (2012) appointed that traditional groups—like establish residents of cities, municipal officials, and homeowners among others—have great hesitancy to involve with the themes of urban expansion. This attitude tends to maintain such perspective obnubilate and somewhat harrowing and impede us to face them in a direct manner.

A particular situation live the city of Arequipa, located in the western foothills of the Andes in Southern Peru. The historic urban center, seated on a slice of the Chili River valley, has a perfect geometry in chess (see Fig. 6.12). The buildings were built with white volcanic rock blocks, called "sillar," originated by the ancient extrusions of three volcanoes in the vicinity (Misti, that is active, and Picchu Picchu and Chachani, inactive). The historic center of Arequipa, or the so-called Ciudad Blanca, is one of the most imposing of South America and is recognized as a cultural heritage site by UNESCO (Fig. 6.13).

When this architectural relic began its construction in 1540, the small village housed no more than 200 people. At that time, if there was a volcanic extrusion, the evacuation of the village would be a reasonable procedure to be put in place. The security of citizens from geohazard could do from the usual procedures for nearby cities to active volcanic apparatus.

In 1940, the city had more than thirty thousand inhabitants and, since then, the population has grown to about nine hundred thousand inhabitants (MPA, Environmental Atlas of Arequipa 2001). All these people that were incorporated into the city are very proud of the architectural beauty and the local volcanic landscape. But, they did not realize the new condition of the current great Arequipa, whose boundaries can no longer be seen from the terraces of the highest buildings.

Fig. 6.12 Historic Centre of Arequipa, declared a World Heritage site by UNESCO, planned as a perfect board. The dark gray square, in the center, identifies the main square (*Photo* By permission of MPA, Environmental Atlas of Arequipa 2000)

Fig. 6.13 Impressive historical buildings in the main square of Arequipa built with white volcanic rock blocks *Credit* Menegat and Fontana

The edge of Arequipa spread out of sight toward the high elevations, just where are the slopes of the Misti volcano, as shown in Fig. 6.14.

Although Misti had a volcanic extrusion long-time ago, it takes part of the very active Andean volcanic arc and seismic zone. Thus, there are risks of avalanches produced by earthquakes and significant torrential flows due to heavy rains. The problem of Arequipa was thus doubled: In addition to geoharzard, there is now also the problem of emergently evacuation of a large city with no plans and infrastructure for procedures of this magnitude. The situation is as if the city had grown by just looking at its center, without considering the possible geological risks surrounding, which at the beginning of the construction of the city seemed far away. The city thus seems no longer realize the geoharzard of the surrounding environment.

The situation of Arequipa is just more didactic than others that we have studied. In general, all cities that have grown a lot in the last decades are in similar situations. In Porto Alegre (Brazil), for example, Guaíba Lake is the only source for water supply. However, on the banks of this lake it is also installed a powerful industry park (petrochemical, pulp, oil refining, tanning, metal mechanics, food, fertilizer, etc.) with very high potential for water contamination. If an industrial accident reached the waters of Guaíba Lake (Menegat and Carraro 2009), like that of 2006, which killed about 100 ton of fish in a lake tributary, the city would be without water for an extended period.

Fig. 6.14 The boundary of the city of Arequipa, about 800 thousand inhabitants, expands to the base of the slopes of the imposing Misti volcano (5,825 m) Credit Menegat

The growth of techno-urbansphere leads it to take areas with increasing geological risks. Its edges are being pushed up on hill slopes of volcanoes or to areas prone to mass movement, or to the flooding lowlands of, deltas, rivers, lakes, and coastal plains (Pelling and Blackburn 2014). The increasingly massive and toxic waste, effluents, and emissions infest the entire region where it is located. In addition to the water contamination perspective, resides the interaction of cities activities and material fluxes of sedimentary and water systems, lead to coastal erosion, changes of sedimentary taxes and river runoff patterns, sequestration of waters in reservoirs, etc. (Crossland et al. 2005). In addition, global changes due to climatic constrains tend to affect lowlands in coastal regions (Dewan 2013). Unlike the idea that cities are safe fortress, the city Mundus can no longer adequately protect its citizens from natural events and also those that, although they have natural scale, were produced as a result of human activity. The researcher Janet Abramovitz (2001) calls these events of "unnatural disasters," including hurricanes and floods arising from climate change.

6.3.3 The Complexity of Techno-Urbansphere Needs Geoethic Culture and Geoknowledge for All Citizens

In general, citizens do not have access to proper representations for understanding the complexity the city where they live (Thierstein and Förste 2008) and much less of the entire techno-urbansphere. The general representation of the city is made by skylines, pictures and simple maps with streets, avenues, and squares. Along the transit by streets and avenues, citizens are captured by the urban mechanical routine and have a very partial vision of the entire city as well as the ecosystems and geosystems that support their vital needs (Pésci 2006). In addition, the physical and cultural city is represented by certain monuments that concentrate all others signs and functions of the urban place. Then, it is enough visiting these monuments to replace city's places. This signal switching or urban semiotic rule assumes the symbol of the city as it were the city itself. The excess semiotic charge in very few city symbols contributes to block the search for others kind of urban representation that could be more appropriate to help citizens understand urban dynamics hierarchies and social and environmental impacts. This would be similar to that situation in which we only classify rocks instead of considering that Earth materials result from the global dynamics of plate tectonics.

The techno-urbansphere is a global system with an extra-human scale, or geological scale, due to magnitude. So when we see the city on the massive scale of techno-urbansphere, there is no resolution to seeing humans, and vice versa. The recognition of its extension is possible only with advanced technological devices (such as satellite images at night, digital geoinformation systems) and, furthermore, scientific theories and education about it are still incipient. Some environmental disasters involving ever larger numbers of people are not understood as a disruption resulting of the inadequate urban interfaces with the entire geosystem where it is embedded, which leads citizens to have growing fear of a "mythical nature" and the city itself. These factors contribute to increase the social disorder and lead to arise old survival strategies pertaining to the natural state of man, including violence and barbarism (Menegat 2006).

Due to lack adequate representations of the city growing, citizens have difficult to update the subjective image of the small and isolated city that they inherited from their parents and predecessor generations. The contradiction between the subjective image of city, where citizens think to be living, and the complexity of the ongoing megacity agglomeration, where they actually are living, is a great challenge to management the urban realm. The contradictions among three successive generations and the correspondent steps of the city growing during the same time interval could be illustrated by broccolis metaphor. The broccoli as cities (Batty and Longley 1994) show fractal geometry and its stems have different sections from base to top (see Fig. 6.15). The basal branch's section is single and circular, and it represents the small and isolated city of the grandparents' generation. In the middle, the single branch forks to three branches and give way to four sections that show the situation of the city growing during the parents' generation. However, at the very top, there are many branches, and the section is composed by a patch of coalescent circles, which form agglomerate that identify the real situation of children generation.

The successional stages of broccoli's branches sections illustrate steps of growing cities from isolated to huge agglomeration. In Brazil, the three stages

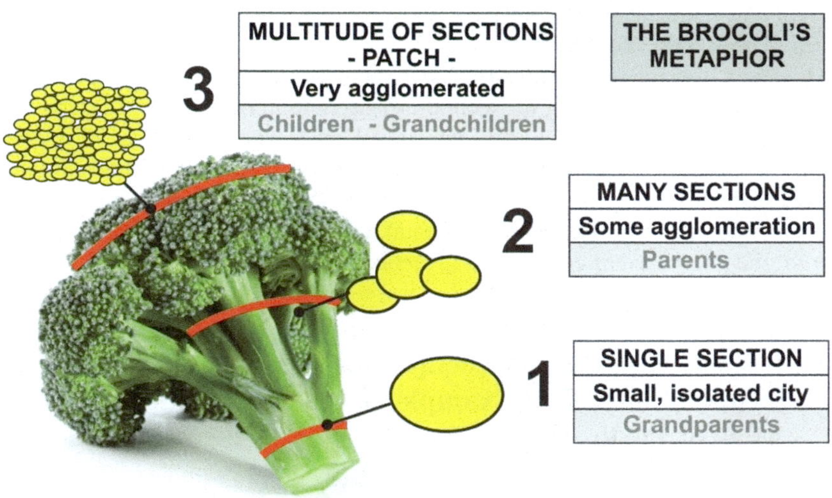

Fig. 6.15 Complex hierarchy of broccoli sections is analogous to successive steps of growing cities. While the base stem is constituted by a single circular Section (1), the top shows a multitude of coalescent circles (3). The isolated circular section at the base is analogous to small city which corresponds to the city of the grandparents times. Now the section with multiple circles at top corresponds to the complexity of the city of the grandchildren times, where citizens actually are living *Credit* Menegat and Fontana

correspond to: (1) metropolitan cities of the 1940s; (2) the urban conurbations explosion of the 1970s; (3) nowadays megacities. In other words, most of the inhabitants of megacities over 30 years old have an inherited culture of their parents and grandparents, who lived in an urban situation represented by stages 1 and 2. The cultural gap worsens for future: in 2050, the children and grandchildren of the current inhabitants of megacities will live, in a world with 9 billion inhabitants, being 6 billion in cities. This huge cultural gap needs the perspective of geological science view, in terms of deep time and geodynamics, and mainly by Geoethics thinking.

The challenges of geosciences for the next decades are immense and perhaps this will be one of the most ordered professional fields. Demand for construction materials for growing cities not only tend to increase as well as new perspectives open up with the search for resources that enable lower environmental impact in their extraction and better environmental comfort of the building, reducing energy expenditure (Hough 1989; Sattler 2004; Oliveira 2014). In addition, the quest for Geoethics and sustainability demands new social responsibility and environment care in the mining industry, which is much known for environmental damage, social deterioration, and illegal trade practices (Irina and Stückelberger 2014; Dashwood 2014). The urban water supply arises as very intricate issue. On the one hand, there is increased demand for water, on the other, the running water close to the cities are increasingly polluted. Furthermore, urbanization has significant impacts on the hydrological cycle, blocking rivers due to the construction of dams either to water supply either to hydroelectricity plants. Megacities tend to change the rain and runoff patterns with the increasing frequency and magnitude of urban floods (Tucci 2006; Zevenbergen et al. 2010).

However, the biggest challenges are related to the knowledge production in describing and understanding the techno-urbansphere in order to help design it and planning to reduce their ecological footprint (Wackernagel and Rees 1996), and its impact on other planetary spheres. Therefore, the information production on urban geolandscape to managers, technicians, scientists, legislators, judges, entrepreneurs, students, and citizens arises as fundamental to build an urban science and culture able to grasp the current complexity (Thierstein and Förste 2008).

6.4 Geoknowledge and Geoethics for Urban Sustainability: The Example of the Environmental Atlas of Porto Alegre City

The Porto Alegre city, on the banks of Guaíba Lake, is the capital of Rio Grande do Sul, the southernmost state of Brazil, and it has a population of about 1.4 million inhabitants. The quality of life in the city can be evaluated by various socio-environmental indicators (see Menegat et al. 2000; Menegat 2002a, b; Menegat et al. 2006): (a) drinking water and sanitation reach, respectively, 100 and 87.7% of the total population (PMPA-DMAE 2015); (b) public streets have more

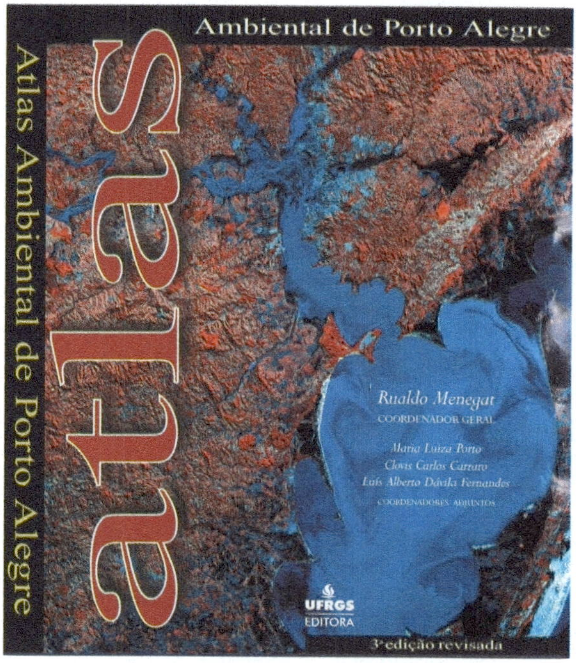

Fig. 6.16 Cover of Environmental Atlas of Porto Alegre, published in 1998, after four years and nine months of work made by a team with over two hundred technicians and researchers *Credit* Menegat and Fontana

than one million trees only along the boulevards and sidewalks, the most wooded capital of the country; (c) the green area index is 14.95 m^2/inhabitant; (d) the population participates in the decision process of investments in the city through the participatory budget system, inaugurated in 1989.

In 1998, the Environmental Atlas of Porto Alegre (Fig. 6.16) was published as a result of an partnership between the Federal University of Rio Grande do Sul (through the Geosciences Institute), the municipality of Porto Alegre (through the Environmental Board of Municipality), and the National Institute for Space Research. This Atlas presents the natural history of Porto Alegre from 800 million years ago (the date of the oldest rocks) to the present time, when take place the intricate mixture of gases in the atmosphere due to vehicle emissions.

The Atlas presents a large amount, superbly documented of data on the environmental parameters of the municipality, as well as models synthesizing data using advanced interdisciplinary environmental approach. From the matrix of the geological basic knowledge (Menegat et al. 1998), it was possible to establish a Natural History of the landscape, as pointed by the great naturalists of the nineteenth century, as Alexander von Humboldt and Charles Darwin, but using modern technologies and theories of science as well as the current worldview based on

Geoethics, citizen and open science, public participation for sustainable cities (Irwin 1995; Fischer 2000; Dickinson and Bonney 2012).

The themes of the natural history of Porto Alegre were organized into three great sections: the Natural System, with eight chapters on geology, geomorphology, hydrology, soils, vegetation, fauna, climate, and conservation of natural areas; the Man-made System with nine chapters that present urban evolution, urban spatial model, the evolution of green areas, urban arborization, urban green areas, urban microclimate, urban activities environmental impact, and sanitation services. Finally, the Environmental Management System, where three chapters present key concepts and problems of environmental management in an urbanized world, public environmental management and general data of Porto Alegre.

6.4.1 Geology and Urban Environment: Understanding the Natural History of the City

The Environmental Atlas provides a complete understanding and representation of the Porto Alegre environment evolution through time, as proposed by geological approach. Besides techniques like thematic maps (two-dimensional representation) and block diagrams (three-dimensional representation) presented in a succession of many scales (from local to regional, continental, and planetary scenarios), it was illustrated in several steps of landscape evolution through geological time (see Fig. 6.17). Several block diagrams illustrate the geological evolution through geological time—since the Precambrian continental collision until Quaternary sea level changes. Furthermore, it also was illustrated the events that occur in the present time scale measured by hours, days, months and years—like the progression of a cold front during two or three days in winter, or a vegetation succession of islands of the Jacuí delta

The field surveys of geological, pedological, and other underground facts as well as the geomorphological, hydrographic, vegetational, faunistic, atmospheric, and many others urbanistic facts of the superficial landscape were showed by 98 thematic maps. These maps were integrated by computerized GIS systems and published in a range of scales, from detailed urban and vegetational maps in 1:10,000 to more general in 1:400,000. In its turn, all thematic units of maps were illustrated through three-dimensional models (block diagrams) and watercolor dioramas and panoramas based on the very detailed integration (vertical and horizontal dimensions of ground and underground continuum) of all scientific data obtained. In addition, diorama, block-diagram, and legend were accompanied by 611 photographs, giving the link between the landscape and how an observer could see it in terms of the concepts represented in the Atlas. The photographical plan also included several scales, from flights with airplanes and helicopters, to shots using cranes in avenues and streets, and finally, microphotography in laboratories. The accuracy of the technical language combined with all these illustrations that were produced in the same project that took place during four years and nine months lead

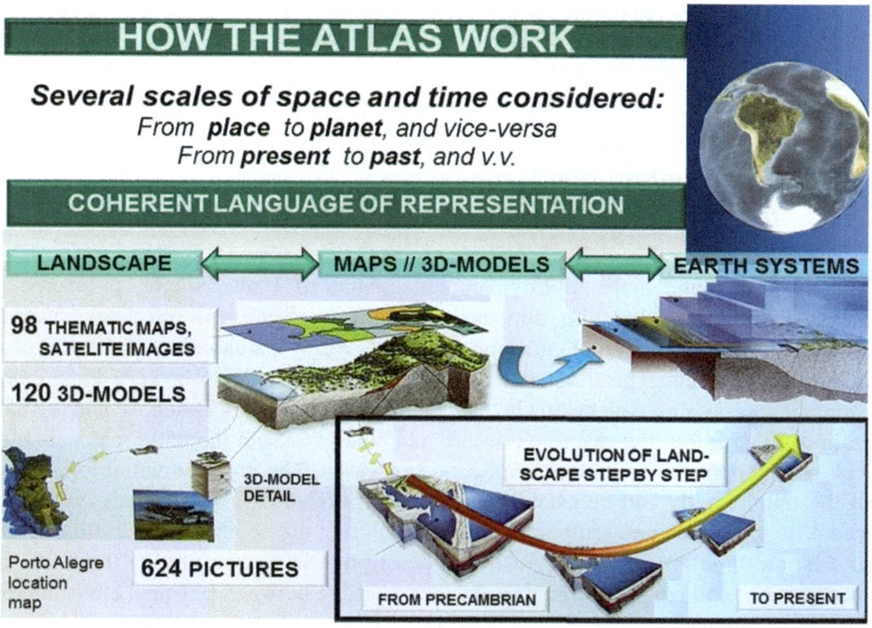

Fig. 6.17 Scheme of dioramas and block diagrams showing the relationship among a succession of scales (from local to planetary) and evolution through geological times (Quaternary sea level changes) used to the explain events represented by thematic maps of the Environmental Atlas of Porto Alegre *Credit* Menegat and Fontana

to a coherent scientific book. The several aspects of the environment become accessible to the public not specialized and technicians from different areas, promoting interdisciplinary and systemic understanding of the subjects dealt with.

The Environmental Atlas of Porto Alegre provided several technical, scientific, and institutional advances grouped into four perspectives: (a) scientific knowledge and interdisciplinary approach among Earth and life sciences, and urban themes; (b) representation of cities, their environmental continuum and evolution trough geological time; (c) partnership among municipality, academies and research institutions; (d) citizen science, environmental and urban management, citizen participation and geoeducation toward an geothic approach.

Currently, nearly 40 cities have developed and published their own atlas based on the reference of the Environmental Atlas of Porto Alegre, including (a) Brazilian cities: Recife (PMR 1999), Joinville (FATMA 2002), São Paulo city (Sepe and Takia 2004), Araraquara (PMA 2004), Brasilia (DF 2006), Salvador (PMS 2006), Londrina (Barros et al. 2007), São José dos Campos 2007 (Morelli 2007), Santana de Parnaíba, SP (PMSP 2010), São Sebastião (Saraceni 2010), Bebedouro (Saraceni and Seibel 2008), Itanhaém (Romagnano and Oliveira 2012); (b) South America cities: Buenos Aires (Nabel and Kullock 2006a, b), Lima (MPL 2008), Trujillo

(MPT 2002) and Arequipa (MPA 2001); and (c) European cities: Barcelona (Acebillo and Folch 2000).

6.4.2 Urban Environmental Management, Citizen Participation and Geoeducation: Geoethics and Paths to Tame the Techno-Urbansphere

The main thematic maps of Environmental Atlas of Porto Alegre, published at 1:100.000 scale (maps with dimensions of 32 cm × 41 cm), provides detailed access to the analysis and crossing-data of several parameters of natural and man-made systems. The diagnosis of environmental accidents or very local environmental problems can be made using adequate technical concepts and visualization. In addition, the macro-environmental analysis from thematic maps enables technical support to numerous inspection processes. The environmental legislation of the municipality can be established based on scenarios and forecasts built on a scientific concepts and empirical data. Similarly, areas of great environmental interest are highlighted in analytical and descriptive terms, improving the forms of management and conservation. The Atlas is a bridge between physical environment and social environment and aid to avoid false polemic triggered by the eco-ideology that often underpins the environmental issues. Appropriately also circumvents up the urban-ideology, which often repulse the analysis of the real importance of environmental issues in the quality of life in urban areas.

On another hand, citizens have access to the best tool for local environmental management: knowledge (see Fig. 6.18). The Atlas gives support to citizens to gain confidence in their ability to manage the environment, going beyond the prescriptions that sometimes do not apply to the immediate reality in which they lives. Environmental education, geoeducation (Silva et al. 2015), and place-based education (Smith and Gruenewald 2007) can be developed in each classroom with the use of local information (Menegat 2000).

There is a renewed success in the perspective of a geoeducation through the Environmental Atlas of Porto Alegre. First, the student motivation which are always more interested when the object of learning part from himself experience. Second, by understanding his reality from environmental assumptions, early the student qualifies himself for the conscious exercise of citizenship (Lynch 1960; Smith and Gruenewald 2007). Thus, environmental education implies in factors that need to be defined in broader scales than those are customarily performed. There is the need for interdisciplinary treatment of the problem—where the geological knowledge arises as fundamental—and assemblage of various institutions, so as to concentrate efforts and predict adequately objectives and results.

Geoeducational programs with these assumptions have been developed in Porto Alegre through the so-called Intelligence Laboratory of Urban Environment, ILUE (Fig. 6.19). This program based on the Environmental Atlas began just after the launch of the book, through a course named as “How use the Atlas in class.”

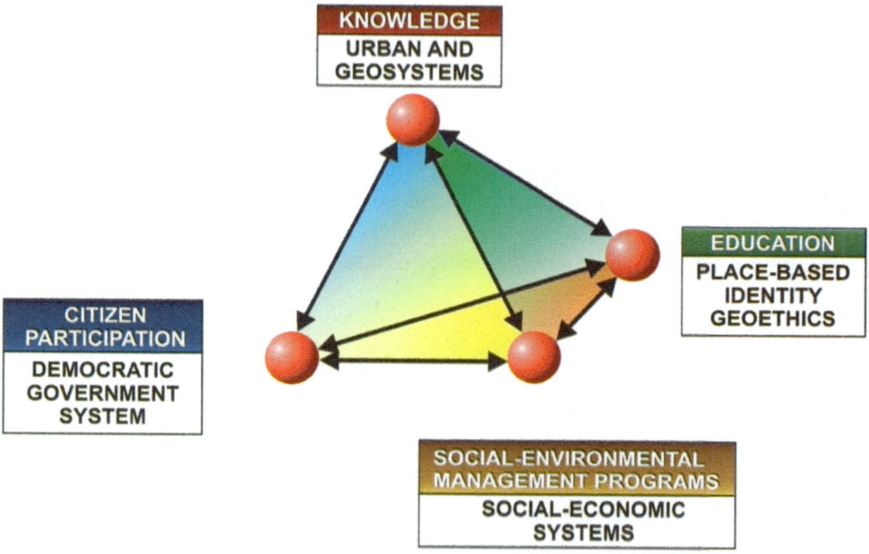

Fig. 6.18 Tetrahedron of the integrated urban geoeducation and Geoethics. The four points of the strategy are the vertices and edges represent mutual interactions. None vertice is most important than another one After Menegat (2000)

Several editions of the course were taught by Atlas' team since 1999 until nowadays, forming more than 400 teachers of municipal schools. This process changed deeply environmental learning and teaching in municipal schools of elementary education, as occurred at Judith Macedo de Araújo School, where the first ILUE was implemented since 1999 (Menegat 2000).

In addition, other decision spheres of city's destiny earn a tool for comprehensive understanding of the reality in which they operate. The Participatory

Fig. 6.19 Intelligence Laboratory of Urban Environment in municipal school of Porto Alegre. In the foreground, the rocks collection assembled by students and school teachers supervised by students of geology of the University *Credit* Menegat and Fontana

Budget (Menegat 2002a, b), an undeniable mark of Porto Alegre, can be enriched as its protagonists can discuss based on scientific representations of urban reality beyond the immediacy of the demands. Based on Environmental Atlas, technicians, students, teachers, managers, leaders, entrepreneurs, non-governmental organizations, and citizens can establish common conceptual framework, improving the logic and efficiency of practical solutions to the problems they face (Fig. 6.18). Environmental management programs can be understood in all the clippings of social and urban life, leaving of the hermeticism of technocratic spheres, or eminently environmental and even academics circles.

6.5 Final Remarks: Toward a Geoethic Culture for Urban Realm

Geology has to offer major contributions to solving urban problems, which are not limited only to classical mineral and energy resources exploitation. There are a number of issues related to environmental management, such as the proper disposal of solid, liquid, and gaseous waste, which has received increasing attention from professionals and researchers in the Earth sciences. Carbon sequestration as a measure to mitigate the impact of CO_2 emissions on climate change is one of them. Or, the searches for new sources of water supply, control of aquifer pollution, urban geomorphology studies are other important fronts. All help to understand the effects of urbanization and reduce the ecological footprint.

However, there are three aspects whose demand for geological knowledge should increase greatly in the coming period. The first refers to the urban planning and reduction of environmental and geological hazard. The urban spread, mainly spontaneous, pushes cities to inhospitable places and high geological and environmental hazard. Besides the use of canonical techniques to reduce risks, it is necessary to establish strategic plans for environmental emergencies, especially in megacities. In Brazil, the major urban centers do not have control mechanisms and unified management. Each municipality cares only about “their part” and the metropolitan control agencies do not have sufficient mechanisms to contribute to effective strategies in emergency situations.

The second challenge is to contribute to the design and investigations of techno-urbansphere by making use of geological techniques like regional scales terrain modeling and medium and long-term forecast. Megalopolis and megacities are physically dimensioned only on the size of its area, population, density, and height of buildings. Although the ecological footprint measurement techniques have brought great advances in the design of urban consumption and waste disposal, we need to scale geological interfaces of the city of the world with the other planetary spheres.

The third aspect concerns the understanding of humanity’s place in nature. It is necessary to develop representations of cities from thematic surveys, photographs and three-dimensional and four-dimensional illustrations that are suitable for interdisciplinary evaluation of urban problems and for formal and non-formal education.

The knowledge about Earth has always been an inseparable part of human progress (Menegat 2008a). In the Renaissance, when there was the discovery of the New World, the mapping of the continents and seas has added a new view of the Earth which has advanced scientific theories about the world we live in. Not coincidentally, in that time took the image of the mythological Atlas changed. Instead of supporting the celestial globe, as advocated by the ancient Greeks, the Renaissance began to represent the Atlas carrying on his back the globe (Fig. 6.20). So there was the discovery that there is a terrestrial world very near to the people, which we need carry it. Each inhabitant has a limit to do it and this limit we call human limit.

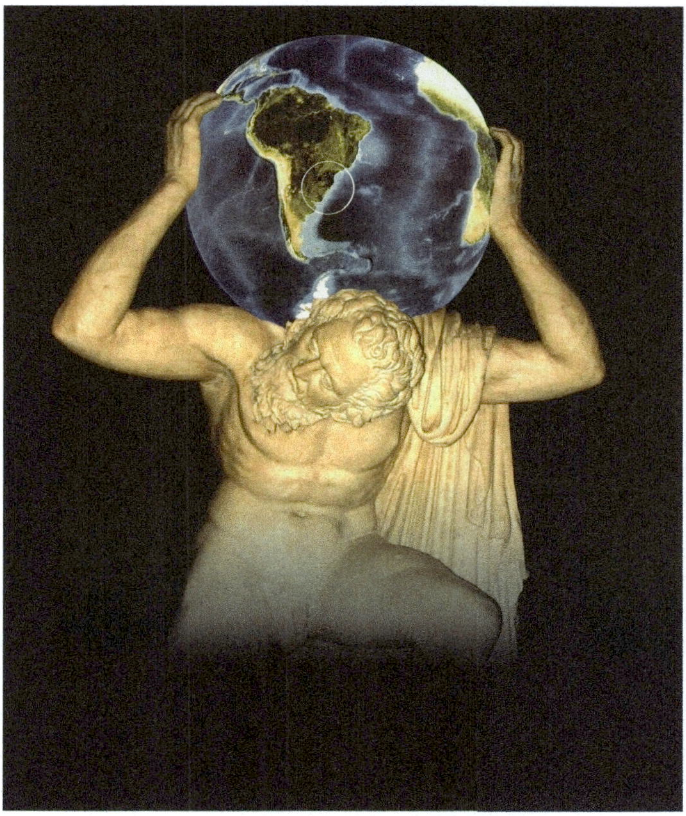

Fig. 6.20 Farnese Atlas carved by Romans in VIII AD. The Renaissance replaced the celestial globe by Earth. In the image above, a spherical projection of the true Earth, obtained by a composite of satellite images has been inserted *Credit* Menegat and Fontana

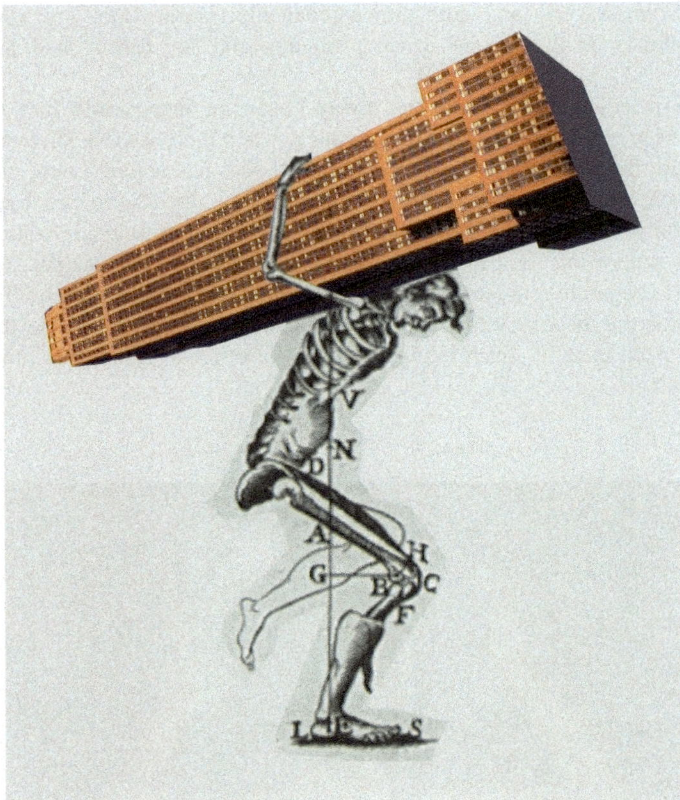

Fig. 6.21 Atlas representation of the XXI century. The Earth was replaced by an immense building, representing the techno-urbansphere, and instead of the mythological god, a human figure was placed displaying their fragile skeleton. This human figure was designed by Giovanni Alphonsi Borelli [1608–1681] and published in his masterpiece De Motu Animalium, where he established the principles of biomechanics *Credit* Menegat and Fontana

To face the urbanized world of the techno-urbansphere, we must, once again, change the representation of the mythological Atlas. If we want that Earth be able to house the biosphere, we must first, make cities more sustainable. Thus, in a more contemporary view, the mythological Atlas should be carrying on his shoulders a building, symbol of techno-urbansphere, and of our primary concerns to reduce the burden of the world (Fig. 6.21).

Acknowledgements We would like to thank to Rogelio D. Acevedo and Jesús Martínez-Frías, general editors, for the kind invitation to take part of this enterprise. Also, to thank to the reviser for many suggestions those improve this chapter. Equally, thanks to Romulo Machado (USP), Celso Dal Ré Carneiro (Unicamp), the Environmental Atlas of Porto Alegre's team, students and teachers of Municipal Schools of Porto Alegre City that construct ILUE day by day. Finally, thanks to CNPq for research support.

References

Abramovitz J (2001) Unnatural disasters. Worldwatch Institute, Washington, p 62
Acebillo J, Folch R (2000) Atles Ambiental de l'àrea de Barcelona; balanç de recursos i problemas. Editorial Ariel, Barcelona, p 439
Adler FR, Tanner CJ (2013) Urban ecosystems: ecological principles for the built environment. Cambridge University Press, Cambridge, UK, p 353
Alasiewicz JZ, Crutzen P, Steffen W (2012) The Anthropocene. In: Gradstein FM, Ogg JG, Schmitz MD, Ogg GM (eds) The geologic time scale 2012. Elsevier, Amsterdam, pp 1033–1040
Angel S (2012) Planet of cities. Lincoln Institute of Land Policy, Cambridge, p 360
Angel S, Parent J, Civco DL, Blei AM (2012) Atlas of urban expansion. Lincoln Institute of Land Policy, Cambridge, p 397
Barker T, Sutcliffe A (1993) Megalopolis: the giant city in history. Palgrave Macmillan, New York, p 228
Bartlett S, Sattethwaite D (eds) (2016) Cities on a finite planet: towards transformative responses to climate change. Earthscan, Routledge, New York, p 306
Barros MVF, Archela RS, Barros OF, Théry H, Mello NA, Gratão LHB (2007) Atlas ambiental da cidade de Londrina. Universidade Estadual de Londrina, Londrina. Available on http://www.uel.br/atlasambiental. Accessed on 10 June 2008
Batty M (2013) The new science of cities. MIT Press, Massachusetts, p 520
Batty M, Longley P (1994) Fractal cities; A geometry of form and function. Academic Press Lim, London, p 394
Bell G, Tyrwhitt J (1972a) Human identity in the urban environment. Penguin Books, Harmondsworth, p 656
Bell G, Tyrwhitt J (1972b). The example of the Tokaido Megalopolis. In: Bell G, Tyrwhitt J (eds) Human identity in the urban environment. Penguin Books, Harmondsworth, pp 533–540
Brinkhoff T (2015) The principal agglomerations of the world. In: Agglomerations of the world, available on http://www.citypopulation.de. Accessed on 15 June 2015
Carson R (1962) Silent Spring. Houghton Mifflin, Boston, p 368
Carvalho,ET de (1999) Geologia urbana para todos: uma visão de Belo Horizonte. ET Carvalho, Belo Horizonte, 175 pp
Chew SC (2001) World ecological degradation; accumulation, urbanization, and deforestation 3000 B.C.–A.D. 2000. AltaMira Press, Walmut Creek, CA, 216 pp
Chrysoulakis N, Castro EA (eds) (2014) Understanding urban metabolism: a tool for Urban Planning. Routledge, London, p 240
Coch NK (2004) Geohazards natural and human. Copley Custom Textbooks, Boston, p 480
Crossland CJ, Kremer HH, Lindeboom HJ, Crossland JIM, Le Tissier MDA (2005) Coastal fluxes in the Anthropocene: the land-ocean interactions in the coastal zone. (Project of the International Geosphere-Biosphere Programme; Global Change—The IGBP Series). Springer, Berlin, p 232
Dashwood HS (2014) The rise of global corporate social responsibility: mining and the spread of global norms. Cambridge University Press, Cambridge, p 336

Dewan A (2013) Floods in a megacity: geospatial techniques in assessing hazards risk and vulnerability. Springer, Berlin, p 199

Distrito Federal DF (2006) Atlas Ambiental do Distrito Federal. Governo do Distrito Federal, Brasília, 56 pp

Dickinson JL, Bonney R (2012) Citizen science: public participation in environmental research. Cornell University Press, Ithaca, p 304

Dogan M, Kasarda JD (eds) (1988a) The metropolis Era: a world of giant cities, vol 1. Sage, London, p 394

Dogan M, Kasarda JD (eds) (1988b) The metropolis Era: mega-cities, vol 2. Sage, London, p 322

Dogan M, Kasarda JD (1988c) Introduction: how giant cities will multiply and grow. In: Dogan M, Kasarda JD (eds) The metropolis Era: a world of giant cities, vol 1. Sage, London, pp 12–29

Dorst JA (1965) Before nature dies. Houghton Mifflin Company, Boston, p 352

Doxiadis CA (1967) The coming world-city: ecumenopolis. In: Toynbee A (ed) Cities of destiny. McGraw-Hill, New York, pp 336–358

Doxiadis CA (1968) Ekistics: an introduction to the science of human settlements. Hutchinson, London, p 527

Doxiadis CA, Papaioannou JG (1974) Ecumenopolis: the inevitable city of the future. Publishing Center, Athens, Athens, p 368

FATMA, Fundação do Meio Ambiente de Santa Catarina (2002) Atlas Ambiental da Região de Joinvile; complexo hídrico da Baía da Babitonga. Fatma: GTZ, Joinvile, 68 pp

Fischer F (2000) Citizens, experts, and the environment: the politics of local knowledge. Duke University Press Books, London, p 352

Fleming F (ed) (1991) The rise of cities. Time-Life Books, New York, 176 pp

Frieling DH (2000) Deltametropolis. In: Vereniging Deltametropool, available on http://www.deltametropool.nl/v1/pages/english/Deltametropolis.php. Accessed on 12 July 2015

Fyfe WS (1991) Our planet observed: the assault by *Homo sapiens*. In: Mungall C, Mclaren DJ (eds) Planet under stress; the challenge of global change. Oxford University Press, Oxford, pp 1–29

Geddes P (1915) Cities in evolution: an introduction to the town planning movement and to the study of civics. Williams and Norgate, London, p 232

Geikie A (1901) The scenery of Scotland viewed in connection with its physical geology. Macmillan, London, p 580

Geikie A (2009[1909]) Darwin as geologist. Cambridge University Press, Cambridge, 100 pp

Gilbert A (ed) (1996) The mega-city in Latin America. United Nations University Press, Tokyo, p 482

Girardet H (1992) The Gaia Atlas of cities; new directions for sustainable urban living. Gaia Books, London, p 191

Glass M (2015) Conflicting spaces of governance in the imagined Great Lakes megaregion. In: Thün G, Velikov K, McTavish D, Ripley C (eds) Infra Eco Logi Urbanism: a project for the Great Lakes Megaregion. Park Books, Zürich, 192 pp

Gottdiener M, Hutchison R (2006) The new urban sociology, 3rd edn. Westview Press, New York, p 432

Gottman J (1961) Megalopolis; The urbanized Northeastern seaboard of United States. Twentieth Century Fund, New York, 810 pp

Gottman J, Harper R (1990) Since Megalopolis: the urban writings of Jean Gottman. University Press, Johns Hopkins, p 304

Goudie A (2000) The human impact on the natural environment, 5th edn. MIT Press, Cambridge, p 511

Grier CH, Feckes M, Bald K (2002) Comparative analysis of the Rhine-Ruhr Metropolitan Region. Düsseldorf Regional Government, Düsseldorf, p 31

GSC, Geological Survey of Canada (2008) Urban geology. GSC, Ottawa. Available on http://gsc.nrcan.gc.ca/urbgeo/index_e.php. Accessed on 10 June 2008

Hardoy JE, Mitlin D, Satterthwaite D (2001) Environmental problems in an urbanizing world. Earthscan, London, p 448

Harrison J, Hoyler M (2015a) Megaregions: globalization's new urban form?. Edward Elgar Pub, Northampton, p 288
Harrison J, Hoyler M (2015b) Megaregions: foundations, frailties, future. *In:* Harrison J, Hoyler M (eds) Megaregions: globalization's new urban form? Edward Elgar Pub., Northampton, pp. 1–28
Herbert S (2005) Charles Darwin Geologist. Cornell University, Cornell, p 512
Hibbert C (1987) Rome: the biography of a city. Penguin, Harmondsworth, p 387
Hodder I (ed) (2013) Substantive technologies at Çatalhöyük: reports from the 2000-2008 seasons. Har. Com, Ankara, p 544
Hough M (1989) City form and natural process. Routledge, New York, p 304
Irina N, Stückelberger C (2014) Mining ethics and sustainability: papers from the World Mining Congress 2013. Globethics.net, Geneva, p 198
Irwin A (1995) Citizen science: a study of people, expertise and sustainable development. Routledge, New York, p 216
Ito T (1980) Tokaido Megalopolis of Japan. GeoJournal 4(3):231–246
Karvounis A (2015) Urban metabolism. In: Chrysoulakis N, Castro EAde, Moors EJ (eds) Understanding urban metabolism: a tool for urban planning. Routledge, London, pp 03–12
King HA, Carpenter B, Wilson N (2007) Hazard city: assignments in applied geology. Prentice-Hall, New York, CD-ROM
Knox P, Florida R (eds) (2014) Atlas of Cities. Princeton University Press, Princeton, p 256
Kraas F, Aggarwal S, Coy M, Mertins G (eds) (2014) Megacities; Our global urban future. Springer, Heidelberg, p 225
Lakshmanan TR, Anderson WP, Song Y (2015) Knowledge Economy in the Megalopolis: Interactions of innovations in transport, information, production and organizations. Routledge, New York, p 228
Legget RF (1973) Cities and geology. McGraw-Hill, New York, p 624
Leveson D (1980) Geology and the urban environment. Oxford University Press, Oxford, p 386
Lietzke B, Vogt R, Young DT, Grimmond CSB (2015) Physical fluxes in the urban environment. In: Chrysoulakis N, Castro EA de, Moors EJ (eds) Understanding urban metabolism: a tool for urban planning. Routledge, London, pp 29–44
Lovelock JE (1979) Gaia: a new look of life on Earth. Oxford University Press, Oxford, p 170
Lowder S (1987) Inside the third world city. Croom Helm, London, p 420
Lüttich H (2010) Metropolitan regions all over the EU. Books on Demand, Norderstedt, p 284
Lynch K (1960) The Image of the City. MIT Press, Massachusetts, p 194
Mayhew S, Penny A (1992) The concise Oxford dictionary of Geography. Oxford University Press, Oxford, p 186
McGill JT (1964) Growing importance of urban geology. United States Geological Survey, Washington, Circular, p 487
Meadows DH, Meadows DL, Randers J, Behrens WW III (1972) The limits to growth. Signet, New York, p 207
Menegat E (2003) Limites do Ocidente; um roteiro para o estudo da crise de formas e conteúdos urbanos. Tese de Doutorado, Programa de Pós-Graduação em Planejamento Urbano e Regional, Universidade Federal do Rio de Janeiro, 229 pp
Menegat M (2006) O olho da barbárie. Expressão Popular, São Paulo, p 351
Menegat R (2000) Educação ambiental integrada: o exemplo do Atlas Ambiental de Porto Alegre. In: A. Krug (org.) Utopia e democracia os inéditos viáveis na educação cidadã. Edufrgs, Porto Alegre, pp 507–522
Menegat R (2002a) Participatory democracy and sustainable development: integrated urban environmental management in Porto Alegre, Brazil. Environ Urbanization 2:181–206
Menegat R (2002b) Participatory democracy in Porto Alegre, Brazil. Participatory Learning and Action (PLA) 44:8–11
Menegat R (2008a) The discovery of the modern Earth by René Descartes: the difficult scientific revolution of the terrestrial spheres. Boletim Geociências Petrobras, V.16, No.2, p. 421–453, may/nov. 2008

Menegat R (ed) (2008b) Visões da Terra, entre deuses e máquinas,qual o lugar da humanidade no mundo em que vivemos. Museu da Ufrgs, Porto Alegre, 96 pp

Menegat R, Carraro CC (2009) Manual para saber por que o Guaíba é um lago. Armazém Digital, Porto Alegre, p 113

Menegat R, Fernandes LAD, Koester E, Scherer CMS (1998) Porto Alegre antes do Homem: evolução geológica. In: Menegat R, Porto ML, Carraro CC, Fernandes LAD (eds) Atlas Ambiental de Porto Alegre. Edufrgs, Porto Alegre, pp 11–20

Menegat R, Porto ML, Carraro CC, Fernandes LAD (eds) (2000) Environmental Atlas of Porto Alegre. Condensed English Version. Edufrgs, Porto Alegre, p 60

Menegat R, Porto ML, Carraro CC, Fernandes LAD (eds) (2006) Atlas Ambiental de Porto Alegre, 3rd edn. Edufrgs, Porto Alegre, p 256

Meuser H (2010) Contaminated urban soils. Springer, Heidelberg, p 318

Mitchell JG, Leen S (2001) Urban Sprawl. National Geographic 200(1):48–73

Montgomery MR, Stren R, Cohen B, Reed HE (eds) (2004) Cities transformed: demographic change and its implications in the developing world (Panel on Urban Population Dynamics). Earthscan, London, p 529

Morelli AF (ed) (2007) Atlas Ambiental de São José dos Campos. Universidade do Vale do Paraíba, São José dos Campos. Available on https://sites.google.com/site/atlasambientalsjc/Home. Accessed on 06 June 2015

MPA, Municipalidad Provincial de Arequipa (2001) Atlas Ambiental de Arequipa. Municipalidad Provincial de Arequipa, Arequipa, 130 pp

MPL, Municipalidad Provincial de Lima (2008) Atlas Ambiental de Lima. Municipalidad Provincial de Lima, Lima, 240 pp

MPT, Municipalidad Provincial de Trujillo (2002) Atlas Ambiental de la Ciudad de Trujillo. Municipalidad Provincial de Trujillo, Trujillo, 121 pp

Mumford L (1938) The culture of cities. Harvest/BJ, New York, p 586

Mumford L (1944) The Condition of Man. Harcourt, Brace and World, New York, 467 pp

Mumford L (1961) The city in history; its origins, its transformations, and its prospects. Harcourt, Brace and World, New York, 657 pp

Mumford L (1964) The Myth of the machine. Harccourt Brace Jovanovich, New York, p 496

Nabel PE, Kullock D (2006a) *Atlas Ambiental de Buenos Aires*. Centro de Investigaciones Geoambientales del Museo Argentino de Ciencias Naturales-CONICET, Buenos Aires, p 111

Nabel PE, Kullock D (2006b). Atlas Ambiental de Buenos Aires. Available on www.atlasdebuenosaires.gov.ar/. Accessed on 15 June 2008

Nagashima C (1972) Megalopolis in Japan. In: Bell G, Tyrwhitt J (eds). Human identity in the urban environment. Penguin Books, Harmondsworth, pp 541–556

National Geographic Society (1992) National geographic atlas of the world, 6th edn. National Geographic Society, Washington, 136 pp

National Geographic Society (1994) Boston to Washington Megalopolis. Map 1:1,000,000

OECD (2015) Urban policy reviews: China 2015. OECD Publishing, Paris, p 243

Okata J, Murayama A (2011) Tokyo's urban growth, urban form and sustainability. In: Sorensen A, Okata J (eds) Megacities: urban form, governance, and sustainability. New York, Springer, pp 15–42

Oliveira R (2014) Optimization of large civil engineering projects from an environmental point of view. In: Lollino G, Arattano M, Giardino M, Oliveira R, Peppoloni S (eds) Engineering Geology for Society and Territory. Springer, New York, pp 01–10

Paccini P, Brunner PH (2012) Metabolism of the Antroposphere. MA, MIT Press, Cambridge, p 392

Papaioannou JG (1964) The city of the future: population projections for Ecumenopolis. Athens Center of Ekistics, Athens, p 186

Park RE, Burgess E, Mckenzie R (1925) The City. University of Chicago Press, Chicago, p 342

Pelling M, Blackburn S (2014) Megacities and the coast; Risk, resilience and transformation. Routledge, London, p 272

Peloggia AUG (1998) O homem e o ambiente geológico; geologia, sociedade e ocupação urbana no Município de São Paulo. Xamã, São Paulo, 1998, p 271
Pésci RO (2006) Vientos verdes: veinte ideas sobre la sustentabilidad. Nobuki, Buenos Aires, p 166
PMA, Prefeitura Municipal de Araquara (2004) Atlas Urbano e Ambiental de Araraquara. Prefeitura Municipal de Araraquara; UFSCAR; UNESP, Araraquara
PMPA-DMAE, Prefeitura Municipal de Porto Alegre. Departamento Municipal de Água e Esgotos (2015) Apresentação. Available on http://www2.portoalegre.rs.gov.br/dmae/default.php?p_secao=318. Accessed on 10 June 2015
PMR, Prefeitura Municipal de Recife (1999) Atlas Ambiental de Recife. Prefeitura Municipal de Recife, Fundo Municipal do Meio Ambiente, Recife, 180 pp
PMS, Prefeitura Municipal de Salvador (2006) Atlas Ambiental infanto-juvenil de Salvador; e diretrizes da educação ambiental. Prefeitura Municipal de Salvador, Salvador, 86 pp
PMSP, Prefeitura Municipal de Santana de Parnaíba. *Atlas Ambiental de Santana de Parnaíba*. Santana de Parnaíba, 190 pp
Reps JW (1998) Bird's eye views; historic lithographs of North American Cities. Princeton Architectural Press, New York, p 116
Romagnano LFT.di, Oliveira RFB (2012) Atlas Ambiental de Santana de Itanhaém. Prefeitura Municipal de Itanhaém, Itanhaém, 130 pp
Sattler MA (2004) Edificações sustentáveis: interface com a natureza do lugar. In: Menegat R, Almeida G (eds). Desenvolvimento sustentável e gestão ambiental nas cidades; estratégias a partir de Porto Alegre. Edufrgs, Porto Alegre, pp 259–288
Saraceni V, Seibel F (eds) (2008) Atlas Ambiental de Bebedouro. Vistadivina, São Paulo, p 108
Saraceni V (ed) (2010) Atlas Ambiental de São Sebastião. Vistadivina, São Paulo, p 115
Sepe PN, Takia H (eds) (2004) Atlas Ambiental de São Paulo. Prefeitura Municipal de Secretaria Municipal do Verde e Meio Ambiente, São Paulo; 266 pp
Serres M (2002) Origins of geometry. Clinamen Press, Manchester, p 256
Short JR (2007) Liquid City: Megalopolis and the Contemporary. Routledge, Washington, p 200
Silva E, Sá AA, Roxo MJ (2015) From Planet Earth to society: a new dynamic in Portugal concerning geoscience education and outreach activities. In: Peppoloni S, Capua G, Di (eds). Geoethics: the role and responsibility of geoscientists. Geological Society, London, pp 141–148
Smith GA, Gruenewald DA (eds) (2007) Place-based education in the global age: local diversity. Routledge, New York, p 408
Syvitski JPM, Vörösmarty CJ, Kettner AJ, Green P (2005) Impact of humans on the flux of terrestrial sediment to the global coastal ocean. Science 308:376–380
Theodorson GA (1982) Urban patterns; studies in human ecology. Pennsylvania State University Press, Philadelphia, p 470
Thierstein A, Förste A (eds) (2008) The image and the region—making Mega-City Regions Visible! Lars Müller Publishers, Baden, Switzerland, 288 pp
Thün G, Velikov K, McTavish D, Ripley C (2015) Infra Eco Logi Urbanism: a project for the Great Lakes Megaregion. Park Books, Zürich, p 192
Toynbee A (1967) Cities in history. In: Toynbee A (ed) 1967. Cities of destiny, McGraw-Hill, New York, pp 11–28
Tucci CEM (2006) Água no meio urbano. In: Rebouças AC, Braga B, Tundisi JG (eds) Águas doces no Brasil; capital ecológico, uso e conservação. Escrituras, São Paulo, pp 399–432
United Nations, Department of Economic and Social Affairs. Population Division (1985) Estimates and projections of urban, rural, and city populations, 1950-2025: the 1982 assessment. UN, New York, 198 pp
United Nations, Department of Economic and Social Affairs. Population Division (1986) Population growth and policies in mega-cities: Seoul. Population Policy, Paper n.4. UN, New York, 43 pp
United Nations, Department of Economic and Social Affairs. Population Division (2007) World population prospects: The 2006 revision; executive summary. UN, New York, 32 pp

United Nations, Department of Economic and Social Affairs. Population Division (2009) World Population Prospects: The 2008 Revision. Working Paper No. ESA/P/WP.210. UN, New York, 32 pp

United Nations, Department of Economic and Social Affairs. Population Division (2015) World Urbanization Prospects: The 2014 Revision. UN, New York, 493 pp

Vai GB (2004) Il testamento di Ulisse Aldrovandi e l'introduzione della parola 'Geologia' nel 1603. In: Vai GB, Cavazza W (eds) Quadricentenario della parola Geologia; Ulisse Aldrovandi, 1603, Bologna. Bologna, Minerva Edizioni, pp 65–110

Vanhaverbeke W (1998) An economic analysis of the Flemish Diamond. Eur Plan Stud 6(4): 425–442

Wackernagel M, Rees W (1996) Our ecological footprint. New Society Pub, Gabriola Island, p 160

Welter VM (2003) Biopolis: Patrick Geddes and the City of Life. MIT Press, Cambridge, p 379

Williams M (1992) Americans and their forests. Cambridge University Press, Cambridge, p 624

Williams M (2002) Deforesting the Earth: from prehistory to global crisis. Chicago University Press, Chicago, p 715

World Bank (2015) East Asia's Changing Urban Landscape: Measuring a Decade of Spatial Growth. Urban Development Series, World Bank, Washington, DC, p 180

Yang J (2009) Spatial planning in Asia: planning and developing megacities and megaregions. In: Ross CL (ed) Megaregions: planning for global competitiveness. Island Press, Washington, DC, pp 35–52

Zellner P, Ruby A (eds) (2000) Southern California: a megalopolis in the making. G. B. Intl, New York, p 340

Zevenbergen C, Cashman A, Evelpidou N, Pasche E, Garvin S, Ashley R (2010) Urban Flood Management. CRC Press, Boca Raton, FL, p 340

Zhang X (2015b) Globalization and the megaregion: investigating the evolution of the Pearl River Delta in a historical perspective. In: Harrison J, Hoyler M (eds) Megaregions: globalization's new urburban form? Edward Elgar Pub., Northampton, pp175–199

Part IV
Chile

Chapter 7
Meteorites and Craters Found in Chile: A Bridge to Introduce the First Attempt for Geoheritage Legal Protection in the Country

Millarca Valenzuela and José Benado

Abstract This contribution presents a short historic review of meteorites and impact craters found in Chile, and how they were recently proposed to be protected as geological heritage, helping to introduce for the first time the concepts of geoheritage and geodiversity to a political level in the country. It includes a table with the percentages of the different types of meteorites found in Chile, for a total number of 1064 official meteorites declared to the Meteoritical Bulletin until July 2017.

Keywords Meteorites · Atacama Desert · Impact craters · Geoheritage

7.1 Introduction

Chile has very particular geological characteristics, only partially shared with other countries: (i) it is on a convergent margin zone, where two oceanic tectonic plates—Nazca and Antarctica—subduct under the South American continental plate. This configuration provokes important seismic activity and the development of a magmatic and tectonic setting that creates one of the largest active volcanic chains in the

M. Valenzuela (✉)
Instituto Milenio de Astrofísica, Pontificia Universidad Católica de Chile PUC, Av. V. Mackenna 4860, Macul, Santiago de Chile, Chile
e-mail: millarca.valenzuela@sernageomin.cl

M. Valenzuela
Servicio Nacional de Geología y Minería, Av. Santa María 0104, Providencia, Santiago de Chile, Chile

M. Valenzuela · J. Benado
Sociedad Geológica de Chile, Valentín Letelier 020 of. 401, Santiago de Chile, Chile

M. Valenzuela · J. Benado
Fundación Centro de Estudios de Montaña, Ramón Carnicer 17, of. 053, Providencia, Santiago de Chile, Chile

R. D. Acevedo and J. M. Frías (eds.), *Geoethics In Latin America*,
The Latin American Studies Book Series,
https://doi.org/10.1007/978-3-319-75373-7_7

world, the Andes Mountains; (ii) it has an abrupt relief which, in less than 200 km, rises from sea level to almost 7000 m above sea level; (iii) it has more than 4300 km of continental coastline; (iv) it presents a latitudinal variation of approximately 38° which is reflected in a great climatic variety, including desertic regions in the north, mediterranean in the center, subpolar and oceanic conditions in the south, and the extreme icy caps present at peninsular and continental Antarctica where Chile has many scientific-military bases.

The characteristics mentioned above, among many others, make Chile a territory with great comparative advantages in terms of geodiversity, and a world-renowned laboratory for the study of the Earth sciences.

One particular territory, the Atacama Desert, in northern Chile, has proved to be an excellent place to understand the effects of hyper-arid climate regimes sustained in a long period of time, as (i) the existence of the most clear skies on Earth, that have attracted the main astronomical observatories of the world to put their telescopes there; (ii) the preservation of enriched mega porphyry copper deposits (Alpers and Brimhall 1988), making the copper industry one of the main economical activities of the country; (iii) the existence of Mars-like soils (Navarro-González et al. 2003) and other extreme environments for the study of extremophile life (Azua-Bustos et al. 2012) as a planetary laboratory for the search of life; (iv) and recently, the discovery of the highest densities of meteorites per km^2 (Hutzler et al. 2016a) in some of its surfaces, due to their old ages and sustained stability over millions of years (Dunai et al. 2005), allowing the preservation of this rare material, as well as some impact structures.

However, at present, the country does not have legislation specifically oriented to preserve its geodiversity, and the need to have an ethical behavior about it is not part of its present social and cultural values.

This paper describes a particular type of geological heritage: meteorites and impact craters found in Chile. It will be given a short historic review of their study in our country, discussing its main values and threats. In addition, it will be shown how these extraterrestrial geological materials and structures have helped to introduce the concepts of geoheritage in the country, allowing the proposal of the first legal request for protection of these materials in the Chilean history.

7.2 Extraterrestrial Rocks and Impact Structures in Chile

Meteorites and micrometeorites (<2 mm) are natural objects that survive their pass through Earth's atmosphere and succeed to arrive to the surface. They fall randomly all over the world, in the order of 10^4–10^5 ton per year (Hutchison 2004), but only less than 1% of this amount has an appropriate mass to be recovered. They can accumulate in desertic areas (Bland et al. 1996) as Antarctica and hot deserts as North West Africa or Atacama (Chile), where the availability of liquid water—the major factor to destroy them—is restricted. Because they can fall everywhere, all countries can deal with the situation to witness a meteorite fall or report a find of these extraterrestrial rocks. However, there is little knowledge about the subject in most countries, and just a few of them have properly addressed the question of how to deal with this natural extraterrestrial material, and conduct protocols for doing so.

The importance of these rocks is given by the fact that they are the oldest solid remnants of the Solar System formation that scientist can access here on Earth. Most of them come from asteroids, but a few can also come from the differentiated crust of the Moon and Mars. Each of them represents a part of the puzzle which scientist from different research areas study in order to better understand how Earth was formed, and how and where life started. So their protection should be considered a must, rather than only a need, because even if it can be of a national interest, it is, above everything, a world heritage natural object.

Today, due to the lack of legal protection in most countries, there is a meteorite market that operates in two modes as a formal and officially ruled market, and another that correspond to a black market. In the first one, most of the offers are private collectors grouped by the International Meteorite Collectors Association (IMCA). It operates guided by an ethical code that rules the behavior of any transaction and respects the law and/or regulations for meteorites in different countries, most of them following the lineaments suggested by the Meteoritical Society.

The Meteoritical Society is an international scientific organization leading and promoting the research on Meteoritics, responsible for the record and authentication of new meteorites through the Meteoritical Bulletin periodic publication. This organization calls to donate at least the 20% of the meteorite to a scientific institution that can perform science with them—the type specimen—, but encouraging the donation of representative samples of meteorites to the local institutions where these meteorites were found, to promote education and research in meteoritics (http://www.meteoriticalsociety.org, accessed August 2017).

Despite the scientific importance of meteorites and impact craters, in Chile they do not have any legal protection. This fact has allowed that a large number of valuable meteorite samples have been extracted from the country, both by scientific expeditions and by private collectors, and that impact crater structures have been affected in their entirety. The causes of this situation are varied and complex, but it is plausible to suppose that the most relevant of them is the lack of knowledge about the scientific importance of these natural elements and of the heritage conservation values in the national community and authorities. In total, there are ~ 190 proven impact structures in the world (http://www.passc.net/EarthImpactDatabase/, accessed August 2017), eleven of them located in South America, being one of them the Monturaqui impact crater, located in northern Chile, that is presented here, as well.

7.2.1 Meteorites Found in Chile

Twelve years ago, Chile had only ~ 60 official meteorites (Muñoz et al. 2007), some of them found by miners looking for ore deposits in the Atacama Desert from the early XIX century until the 1990s, as the historic meteorites Imilac, a pallasite found around 1820 and described in the treatise "Viaje al desierto de Atacama" (Phillippi 1860), and the mesosiderite Vaca Muerta, discovered in 1861, one of the more massive meteorites found in Chile so far, with 3.8 ton, from which Domeyko and Sandt wrote reports (Pedersen et al. 1992). In 1983, private collectors—the brothers Rodrigo and Edmundo Martínez—restarted the search for meteorites in the Atacama Desert, rediscovering big mass fragments from Vaca Muerta and Imilac, and little craters associated to them (www.museodelmeteorito.cl, accessed July 2017). They defined the first dense collection area (DCA) called Pampa (Zolensky et al. 1995), located in the flatlands of Mejillones Península, near the city of Antofagasta, with $\sim$30 meteorites from different falls, collected in an area of 30×15 km^2 (Pampa and La Yesera meteorites), now renamed as Pampa de Mejillones and La Yesera DCAs.

Until those years, Chile was still considered as a place with low potential to find more meteorites due to the difficulty given by the fact that most of the surface rocks were volcanic in origin, looking very alike the stony meteorites (Scherer and Delisle 1992), very different to the perfect contrast displayed in other deserts, as Sahara and Nullarbor.

Other meteorite collections were at the Mineralogical Museum of Copiapó (at least six different meteorites, most of them irons), the Geological Museum Humberto Fuenzalida at Universidad Católica del Norte (Antofagasta), Geological Museum of the Servicio Nacional de Geología y Minería (SERNAGEOMIN, Santiago), and Chilean Natural History Museum (Santiago). Some important samples were robbed from these museums, other were not preserved in the right conditions, forming rust and breaking apart, and other just disappear for lack of a proper procedure to keep them well cataloged. Most of these issues could have been avoided with the good practices of a professional curator.

Little was done scientifically those years, and most geologists took it only as a hobby. Carlos Roeschmann, once the director of the SERNAGEOMIN Laboratory, was one of the only Chilean geologists who were interested to do research on meteorites, making connections with the national private collectors and other international museums, and teaching an elective course of Meteorites in the Geology Department at the Universidad de Chile in the years close to 2000.

Thanks to the development of the first Chilean Ph.D. thesis studying the meteorites in the Atacama Desert (Valenzuela 2011), and the interest that arises in the international scientific community, a collaboration with French partners (CEREGE, Université Aix-Marseille) was possible. Among one of the most important research activities of the Chilean-French team are the annual expeditions to search for meteorites in the Atacama Desert, starting in 2004. Some studies include Valenzuela et al. (2008) who describe and quantify the weathering products found in Atacama ordinary chondrites (OCs) for the first time, using Mössbauer spectroscopy, while Munayco et al. (2013) continue the characterization of the weathering mineralogy for new samples discovered in other DCAs. The study of the flux of meteorites was addressed in Gattacceca et al. (2011) and Hutzler et al. (2016a), where densities of meteorite per km^2 were calculated for San Juan and El Medano DCAs, respectively, demonstrating an amazing concentration in certain surfaces, as at El Medano DCA, with more than 150 meteorites/km^2, due to the old and stable surfaces of the desert and the existence of meteorites with terrestrial ages reaching 2 My (Hutzler et al. 2016b).

Knowing that a few private collectors were operating in a permanent way in the Atacama desert, most of them working under the procedures established by the IMCA and the Meteoritical Society, the scientific team responsible of the Atacama Meteorite Search Expeditions—lead by Gattacceca and Valenzuela—contact them to offer the possibility of a collaboration. The information about most recoveries in Chilean territory has improved thanks to the help of private collectors. A percentage of each meteorite recovered by this way has been stored in Chilean and/or French collaborating institutions, and the information of important findings has been shared, as the discovery of a new strewnfield of tektites (aerodynamic shaped glasses formed in big impact events) called Atacamaites (Devouard et al. 2014). By their side, they could save time, having an authenticated official meteorite in approximately six months, in contrast with the year they had to wait when they send their samples to American institutions.

At present, the number of official Chilean meteorites is 1064 (https://www.lpi.usra.edu/meteor/metbull.php, accessed in July 2017), and is dominated by OCs by far, corresponding to the 92% of the total amount (Table 7.1). As pointed by Gattacceca et al. (2011), the recovery of other kind of stony meteorites in Atacama, beside OCs, is quite difficult as the surfaces are dominated by volcanic rocks of different types, and without the fusion crust, an achondrite, for example, will appear very similar to the most abundant rocks in the desert.

As is shown in Table 7.1, there are some scarcer meteorite types in the Chilean collection, as carbonaceous chondrites (18), the meteorite El Medano 301 that corresponds to an ungrouped reduced chondrite (Pourkhorsandi et al. 2017), and 46 irons.

7.2.2 *Craters*

Along with research studies relevant to the previously described meteorites in Chile, there have also been reports and studies of extraterrestrial impact craters. The Vaca Muerta set of holes, the Imilac small crater, and the Monturaqui explosion crater are examples of such cases, all of which are located within the Atacama Desert.

The Monturaqui impact crater is a paradigmatic case in terms of conservation because it presents a set of antecedents (Table 7.2) that, according to the Chilean Geological Society (SGCH), elevate it to the category of geological heritage of national relevance (http://sociedadgeologica.cl/crater-de-impacto-monturaqui/, accessed in July 2017). However, it does not count with any legal standing that may protect it, although there are a number of present and potential factors that threaten to damage it (Table 7.2).

The crater is approximately 3000 m above sea level and is located in the Andean foothills, south of the Atacama Salt Flat. It has an almost circular structure, ~350 m in diameter and ~34 m deep (Fig. 7.1). The target rocks of the

Table 7.1 Number of meteorites found in Chile, per types, present in the official list of the Meteoritical Bulletin (July 2017), compared with the total amount of meteorites (total falls and finds of the world), and their percentages

Meteorites	Chile	%[a]	%[b]	Total[c]	%[a]
Stones					
Chondrites					
Carbonaceous	**18**	**0**	**2**	**2162**	**4**
CB	1	0	0	20	0
CH	0	0	0	24	0
CI	0	0	0	9	0
CK	1	0	0	346	1
CM	0	0	0	545	1
CO	10	0	1	558	1
CR	3	0	0	178	0
CV	0	0	0	426	1
Ungrouped	3	0	0	43	0
Ordinary	**981**	**2**	**92**	**49015**	**83**
Petrologic type 3	67	0	6	2809	5
H	547	1	51	22157	38
L	383	1	36	19614	33
LL	51	0	5	7244	12
Other chondrites	0	0	0	1113	2
Enstatite	0	0	0	599	1
Rumuruti	0	0	0	183	0
Kakangari	0	0	0	4	0
Ungrouped chondrites	1	0	0	19	0
Melt (impact)	4	0	0	308	1
Total chondrites	**1004**	**2**	**94**	**52290**	**89**
Achondrites					
Primitives achondrites					
Acapulcoites-lodranites	1	0	0	148	0
Brachinites	0	0	0	41	0
Winonaites	0	0	0	30	0
Subtotal	**1**	**0**	**0**	**219**	**0**
Differentiated achondrites					
Aubrites	0	0	0	74	0
Ureilites	3	0	0	451	1
HEDs	4	0	0	1893	3
Howardites	0	0	0	341	1
Eucrites	3	0	0	1112	2
Diogenites	1	0	0	440	1
Lunar meteorites	0	0	0	306	1
Martian meteorites	0	0	0	198	0
Angrites	0	0	0	28	0

(continued)

Table 7.1 (continued)

Meteorites	Chile	%[a]	%[b]	Total[c]	%[a]
Ungrouped	0	0	0	80	0
Enstatite achondrite				13	0
Subtotal	**7**	**0**	**1**	**3043**	**5**
Total achondrites	**8**	**0**	**1**	**3262**	**6**
Unclassified stony	0	0	0	1888	3
Total stones	**1012**	**2**	**95**	**57440**	**97**
Stony Irons					
Pallasites	1	0	0	114	0
Mesosiderites	5	0	0	235	0
Total stony irons	**6**	**0**	**1**	**349**	**1**
Irons					
Magmatic					
IC	0	0	0	12	0
IIAB	6	0	1	129	0
IIC	0	0	0	8	0
IID	1	0	0	24	0
IIF	0	0	0	6	0
IIG	2	0	0	6	0
IIIAB	16	0	2	309	1
IIIE	1	0	0	18	0
IIIF	1	0	0	9	0
IVA	5	0	0	84	0
IVB	3	0	0	15	0
Subtotal	**35**	**0**	**3**	**620**	**1**
Non-magmatic					
IAB complex	9	0	1	300	1
IIE	1	0	0	22	0
Subtotal	**10**	**0**	**1**	**322**	**1**
Iron-ungrouped	1	0	0	121	0
Iron (not classified further)				97	0
Total irons	**46**	**0**	**4**	**1160**	**2**
Total meteorites	**1064**	**2**	**100**	**58949**	**100**

[a]Percentage calculated from the same category
[b]Percentage from the total amount of meteorites
[c]Total = world falls + finds

impact are composed of Ordovician granite basement cut by various small mafic dikes (1–2 m wide) and covered by a thin (0–5 m) discontinuous layer of Pliocene ignimbrite (Rathbun et al. 2016). Along with the structure, some fragments of the remains of the impacting meteorite were found, a thick group I octahedrite as determined by Buchwald (1975), as well as many impactites.

Table 7.2 Values and threats of the Monturaqui impact crater

Type	Description
Values	
Scientific	It is the only confirmed impact structure from explosion of an asteroid in Chile and one of the few recognized in South America. It is a simple crater (<1 km in diameter), relatively well conserved for its age and, as such, a one of a kind natural laboratory for the morphological study of these type of craters and their processes
Cultural	It posses an exceptional historical value since it was part of the Qhapaq Ñan or "Inca Trail" (Hyslop 1984), which was recognized in 2014 as a World Heritage Site by the UN
Scenic	It has a great scenic value (Fig. 7.1), being part of the Atacama Desert and the Andean Precordillera, offering a privileged view of the Andean massif to the East and the Atacama Salt Flat, to the North. The gullies next to the crater, also, grow a variety of native high Andean plateau flora adding beauty to its surroundings
Threats	
Legal	Chilean law has not yet legislated on any sort of specific protection for impact craters
Collecting	Collection and sales of impactites and specimens from the crater are common and have been performed by both touristic and scientific expeditions. An iconic example of this threat occurred when the television show Meteorite Men, from the Science Channel, visited and camped inside the crater, collected and later sold the specimens they had gathered[a]
Erosion	Erosion inside the crater is common due to the flux of vehicles that enter and exit it
Mining	The entire area is surrounded by mining concessions for both exploration and exploitation purposes

[a]The Monturaqui episode is available at: https://www.youtube.com/watch?v=giZU9mlEMAM

Fig. 7.1 Monturaqui impact crater. View to the North.
Credit Hernan Ugalde (in Ugalde et al. 2007)

Based on the paleomagnetic measurements on rocks inside the crater and the use of cosmogenic isotopes to constrain the terrestrial age of the impactor meteorite, the age of the crater was estimated to be 500–600 ky (Valenzuela et al. 2008). Through microscopic study of the minerals that composed the impactites, Roeschmann and Rada (2000) estimated that the heating effect produced during the collision exceeded 1400 °C and dissipated an amount of energy similar to the explosive impact produced by $\sim$2.2 atomic bombs such as those used in Hiroshima.

The crater has been visited extensively along the years by scientific and non-scientific expeditions, both of them damaging the edges of the crater due to the entrance of motor vehicles inside the structure. The place has been cleaned out of all the remaining impactites and iron shales, but its shape and ejecta deposits still remain, along with its scenic beauty, making it a special place to come to observe and learn about one of the most important geological processes occurring in the Solar System, craterization, and its consequences in the origin and evolution of life on Earth.

Efforts have been done to raise concern about this situation among the atacameñan community of Peine that own these lands. At present SERNAGEOMIN, the SGCH, and the Centro de Estudios de Montaña foundation, along with the Peine community, are trying to declare the Monturaqui crater as a Historic Monument to protect it, presenting the first formal request to the Chilean Government during 2017.

7.3 Law Proposal for Meteorite Protection

It has been a long way to build Meteoritics in Chile. The steps have included a continuous effort to participate in the annual expeditions organized by the Centre Européen de Recherche et d'Enseignement des Géosciences de l'Environnement (CEREGE, France), lead by Dr. Jérôme Gattacceca, in a frame project associated to the study of hot desert meteorites, not only in Chile, but also at Algeria, Tunisia, Morocco, and Iran deserts.

The number of Chilean scientists working in meteorites over time has been almost none, but the actions taken by Valenzuela in different national scientific meetings (Valenzuela 2015, 2016) to raise awareness about the situation of Chilean meteorites, specially after the publication of the high densities discovered at Atacama (Hutzler et al. 2016a), have made a positive response in the geological community.

The main collection of meteorites recovered from Atacama in the last expeditions (2004–2016) is in the French repository of the Group of Planetology, Inner Earth, and Meteorites at CEREGE, in charge of the curator Dr. Jérôme Gattacceca. From 2016, a national repository with a few number of meteorites from expeditions was created at the Astro Engineering Center of the Universidad Católica de Chile (AIUC-PUC) in Santiago, now relocated at SERNAGEOMIN. A private collector, Dr. Holger Pedersen, donated pieces of the historic Chilean meteorites Vaca Muerta and Imilac. Some of them are on a permanent display at AIUC, along with other ones collected in Chile. Figure 7.2 shows the detail of two ordinary chondrites from Atacama displayed there.

During 2017, the International Astronomical Union gave the name Millarca to the main belt asteroid N° 11819, as a recognition to the Chilean geologist Millarca Valenzuela for her work done with the meteorites from Atacama Desert, the development of the Chilean Allsky Camera Network for Astro-geosciences

Fig. 7.2 Detail of two ordinary chondrites shown at the permanent exhibition of meteorites and impact materials at the Astro Engineering Center of the Pontificia Universidad Católica de Chile, Santiago. The samples show preserved fusion crust. Shiny metal grains of kamacite as well as dark shock veins are visible in the polished cut surface shown in the image to the right.

(CHACANA) project (Valenzuela et al. 2017), and for her efforts to raise Meteoritics and Planetary Sciences in the country and South America.

The award had great media coverage that helped to install the subject of Meteoritics and Planetary Sciences in the public domain. As a consequence, the Senator Guido Girardi, part of the governmental initiative called Congreso Futuro, decided to present a proposal for the meteorite protection.

The proposal promotes the protection of meteorites for scientific research, norming the exploration and eventual exploitation of meteorites in the country. The law does not consider penalize the use of the material for commercial purposes as tourism and/or jewelry, if they are declared, and norm that at least 20% of each rock must be donated to the institution that will act as the national repository.

Some important points that the law will consider include:

(1) Any meteorite found in Chile will be declared as a geological heritage natural object.
(2) The proposed public institution for the implementation of the protocols established by the law is the SERNAGEOMIN. This institution will hold the official meteorite repository as well.
(3) The protocol will include the requirement of an official permission to conduct search of meteorites, for any kind of national or foreign missions, as well as for the possession of them. There will be fines for people found with meteorites without the correspondent permission.
(4) Illicit traffic will be penalized with prison and fines.
(5) For meteorites of a special value (for its rarity or size), the donation to the official repository will be of 50% of each rock.
(6) The research done with the Chilean meteorites will have to be communicated before 2 years after the findings.

7.4 Final Considerations

The Chilean Geological Society, through its Geoheritage Group of Specialist, has raised the list of sites considered of geological relevance—known as geosites—to ~70 by now (http://sociedadgeologica.cl/category/geositios/, accessed in July 2017), and has presented to the national authorities a proposal to include the concepts of geodiversity and geological heritage in the context of a new law that proposes the creation of the Service of Biodiversity and Protected Areas, and the National System of Protected Areas of the Ministry of Environment, that will rule and norm the relationship with natural environments in Chile. The first attempt did not succeed, but the reference to the same concepts in the law for the protection of meteorites will open a new way to reach the authorities to try in the future to include them.

The method to create consciousness about this subject in the country is to improve and extend the interactions with the community at different levels. For that, the authors think the Chilean geological community—including students, professionals and scientist—should incorporate into their normal activities the ones related with outreach and education, giving them an equal importance as other professional and academic practices, as it is the only way to create a proper culture of respect and responsibility about nature.

In this context, the projects that encourage the creation of geoparks as Kütralcura (Schilling 2009), in the Araucania region, and the miner geopark of the Bio-Bio's litoral, among others, are excellent initiatives to promote the protection of geological, cultural, and historic heritage, as well as contributing to social, cultural, and economic development of the territory where they are emplaced.

The XV Chilean Geological Congress, which will be held in Concepción in November 2018, will include a session dedicated to discuss this and other issues related to any of the ethical aspects related to our professional practice, to trace the next steps to conquer in our community.

References

Alpers CN, Brimhall GH (1988) Middle Miocene climatic change in the Atacama Desert, northern Chile: evidence from supergene mineralization at La Escondida. Geol Soc Am Bull 100 (10):1640–1656

Azua-Bustos A, Urrejola C, Vicuña R (2012) Life at the dry edge: microorganisms of the Atacama Desert. FEBS Lett 586(18):2939–2945

Bland PA, Berry FJ, Smith TB, Skinner SJ, Pillinger CT (1996) The flux of meteorites to the Earth and weathering in hot desert ordinary chondrite finds. Geochimica et Cosmochimica Acta 60 (11):2053–2059

Buchwald VF (1975) Handbook of iron meteorites: their history, distribution, composition, and structure, 3 vol. University of California Press, Berkeley

Devouard B, Rochette P, Gattacceca J, Barrat JA, Moustard F, Valenzuela EM, Alard O, Balestrieri ML, Bigazzi G, Dos Santos E, Gounelle M, Jambon A, Laridhi-Ouazza N,

Shuster DL, Warner M, Warner M (2014) A new Tektite Strewnfield in Atacama, Chile. In: Abstract #5394 for the 77th annual meeting of the Meteoritical society, Casablanca, Morocco

Dunai TJ, López GAG, Juez-Larré J (2005) Oligocene-Miocene age of aridity in the Atacama Desert revealed by exposure dating of erosion-sensitive landforms. Geology 33(4):321–324

Gattacceca J, Valenzuela M, Uehara M, Jull AJT, Giscard M, Rochette P, Braucher R, Suavet C, Gounelle M, Morata D, Munayco P, Bourot-Denise M, Bourles D, Demory F (2011) The densest meteorite collection area in hot deserts: the San Juan meteorite field (Atacama Desert, Chile). Meteorit Planet Sci 46:1276–1287

Hutchison R (2004) Meteorites: a petrologic, chemical and isotopic synthesis. Cambridge University Press, Cambridge

Hutzler A, Gattacceca J, Rochette P, Braucher R, Carro B, Christensen EJ et al (2016a) Description of a very dense meteorite collection area in western Atacama: insight into the long-term composition of the meteorite flux to Earth. Meteorit Planet Sci 51(3):468–482

Hutzler A, Rochette P, Bourlés D, Gattacceca J, Mercel S, Jull AJT, Valenzuela M (2016b) New insights in preservation of meteorites in hot deserts: the oldest hot desert meteorite collection. In: Abstract #6159, 79th annual meeting of the Meteoritical society August 07–12, 2015, Berlin, Germany

Hyslop J (1984) The inca road system (studies in archaeology). Academic Press, St Louis, Missouri, p 337

Munayco P, Munayco J, Avillez RR, Valenzuela M, Rochette P, Gattacceca J, Scorzelli RB (2013) Weathering of ordinary chondrites from the Atacama Desert, Chile, by Mössbauer spectroscopy and synchrotron radiation X-ray diffraction. Meteorit Planet Sci 48(3):457–473

Munoz C, Guerra N, Martinez-Frias J, Lunar R, Cerda J (2007) The Atacama Desert: a preferential arid region for the recovery of meteorites—find location features and strewnfield distribution patterns. J Arid Environ 71(2):188–200

Navarro-González R, Rainey FA, Molina P, Bagaley DR, Hollen BJ, de la Rosa J et al (2003) Mars-like soils in the Atacama Desert, Chile, and the dry limit of microbial life. Science 302 (5647):1018–1021

Pedersen H, Canut de Bon C, Lindgren H (1992) Vaca Muerta mesosiderite strewnfield. Meteoritics 27:126–135

Philippi RA (1860) Viage al desierto de Atacama hecho de orden del Gobierno de Chile en el verano 1853–54, Librería de E. Anton, 280p

Pourkhorsandi H, Gattacceca J, Devouard B, D'Orazio M, Rochette P, Beck P et al (2017) The ungrouped chondrite El Médano 301 and its comparison with other reduced ordinary chondrites. Geochimica et Cosmochimica Acta 218: 98-113.

Rathbun K, Ukstins I, Drop S, Gutierrez F (2016) A new geologic map of Monturaqui meteorite impact crater, Chile: implications for satellite-based geologic mapping of small craters. In: 47th lunar and planetary science conference. 2p

Roeschmann C, Rada C (2000) El Impacto meteórico de Monturaqui, región de Antofagasta, Chile: Productos y Procesos. En actas del IX Congreso Geológico Chileno, vol I, Puerto Varas, Chile, pp 681–686

Scherer P, Delisle G (1992) Are there high meteorite concentrations in the Atacama Desert, Chile? Meteoritics 27:285–286

Schilling M (2009) Hacia la creación del primer geoparque en Chile: Parque Nacional Conguillío, Región de la Araucanía. XII Congreso Geológico Chileno, Simposio de Geopatrimonio, Santiago, 22–26 noviembre, artículo S5_027, 4p

Ugalde H, Valenzuela M, Milkereit B (2007) An integrated geophysical and geological study of the Monturaqui impact crater, Chile. Meteorit Planet Sci 42(12):2153–2163

Valenzuela M (2011) Procesos de meteorización en condritos ordinarios del Desierto de Atacama, Norte de Chile: nuevos antecedentes sobre meteorización de material extraterrestre en ambientes desérticos Tesis para optar al grado académico de Ph.D. Inédito. Departamento de Geología, Universidad de Chile, Santiago, Chile. 201p

Valenzuela M (2016) Chilean meteorites: their heritage status and present research. In: XIII annual SOCHIAS meeting, Antofagasta, Chile

Valenzuela M, Bourles DL, Braucher R, Faestermann T, Finkelb RC, Gattacceca J et al (2008). New age estimation of the Monturaqui Impact Crater. In: 2008 annual report of the Tandem Accelerator Facility, University of Munich, Germany, 27

Valenzuela M, Baeza L, Vanzi L (2015) Meteoritos chilenos: resultados obtenidos luego de diez años de búsqueda en el Desierto de Atacama. Resumen expandido para el XIV Congreso Geológico Chileno, La Serena, Chile

Valenzuela M, Ropert S, Suc V, Jordan A, Vanzi L, Aravena S (2017) CHACANA: Chilean Allsky camera network for Astro-geosciences. In: Abstract for the asteroids, comets, meteors conference—ACM2017, Montevideo, Uruguay

Zolensky ME, Martinez R, Martinez E (1995) New L chondrites from the Atacama Desert, Chile. Meteoritics 30:785

Part V
Colombia

Chapter 8
Geology as a Profession and Geoethics in Colombia

Michel Hermelin Arbaux and Orlando Navas

Abstract At the time America was discovered, three activities related to Earth Sciences were carried out by the native Indians (Amerindians would be a better term) in what is now the Colombian territory: gold and emerald mining and salt extraction; the first one is recognized as the main motivation for Spaniards to conquer the entire mainland and to oblige Indians to work in gold mines; this servitude, together with the new diseases brought from Europe and the occupation of croplands by cattle and horse raising, caused the reduction of indigenous people in 90%, as it was demonstrated by historians as Colmenares (Historia económica y social de Colombia, 1537–1719. Universidad del Valle, Cali, 1973). This situation motivated Bishop Bartolomé de las Casas to obtain from the Council of Indies to bring of African slaves, who were more resistant than Indians to work in the gold mines. The main argument behind was de las Casas conviction that Indians had a soul and thus should be considered as "true human beings." This episode, which signified the forceful translation of hundreds of thousands of Africans to America, should deserve a mention in a topic as "Geoethics."

Keywords Geology and geoethics in Colombia · History

Prof. Michel Hermelin Arbaux diseased

O. Navas (✉)
Consejo Profesional de Geología, CPG, Diagonal 53 # 34 - 53 Of.: 138, Bogotá, Colombia
e-mail: onavas2011@gmail.com

M. H. Arbaux
Bogotá, Colombia

R. D. Acevedo and J. M. Frías (eds.), *Geoethics In Latin America*, The Latin American Studies Book Series,
https://doi.org/10.1007/978-3-319-75373-7_8

8.1 Introduction

During colonial times, very little attention was paid to Earth Sciences: Gold but also silver and emeralds were extracted using very primitive methods (West 1972). Earthquakes and volcanic eruptions were poorly described and only recorded as "acts of God."

Nineteenth century began with the visit of Humboldt, who left a wide report on mineralogy, geology, and botany. It was followed by several Europeans who lived or traveled extensively in the country and left descriptions of interest for the Earth Sciences: Boussingault (1824–1830), Hettner (1882–1884), von Schenck (1880), and Reiss and Stübel (1868–1869) among others. The first Colombian who received formal geological training was J. Acosta (1800–1852) who studied in Paris under de Beaumont and left several writings. The discovery and preliminary reports of the Guajira coal deposits were made by Jorge Isaac (1837–1895) who survived in the memory of the country more as a writer than as a geologist.

8.2 National Institutions Related to Earth Sciences in Colombia

The "Comision Corográfica" (Chorographic Commission) was created under the direction of A. Codazzi, an Italian military engineer and geographer. Unfortunately, the publication of the results was delayed due to civil wars and if they were considered as important geographical documents, their contribution to Earth Sciences is quite limited.

In 1919, the Comisión Científica Nacional was created, and H. Scheibe, a German geologist, was hired as its first director in order to initiate geological studies in Colombia. It was replaced in 1938 by the Servicio Geológico Nacional (National Geological Survey). The Ministry of Mines and Petroleum later replaced by the Ministry of Mines and Energy started its activities in 1940. Ecopetrol, the national petroleum company, was founded in 1951.

Ingeominas started operating in 1969 its activities, which comprised these SGN and Inventario Minero, the later created in 1963. Ingeominas suffered several reforms until in 2013 it became the Servicio Geológico Colombiano (Colombian Geological Survey).

A private entity created by the Pontificia Universidad Javeriana (Jesuit University) under the direction of J. E. Ramírez, the Instituto Geofísico de los Andes Colombianos (Geophysical Institute of the Colombian Andes) was founded in 1941, and until the installation of the national seismic net in the last two decade of last century, it was the only source for seismic information of the country.

8.3 University Programs in Earth Sciences in Colombia

The Medellin School of Mines was founded in this city in 1887 and trained civil and mining engineers. Its professors taught courses and published several important contributions on Geology.

In the 1940 decade, a small group of professors headed by G. Botero decided to create a program in Petroleum and Geology Engineering, which started in 1942. It was split in 1966, and Geology Engineering is still taught at the School of Mines now Facultad de Minas, a division of the National University of Colombia. In Bogotá, the same Universidad National started in 1956, a program in Geology and Geophysics, which later became a Geology program. Recently, Geophysics has been offered as a graduate option. During the following decades, several programs were created both for undergraduate and graduate levels (Table 8.1).

The number of graduates in Geology in Colombia has been increasing steadily since the 1980 decade, when most of the new schools were founded (Fig. 8.1).

A first conclusion can be drawn from the previous information: During many decades, institutions as the Geological Survey and the Ministry of Mines had to rely on geologists trained abroad, mainly foreigners. Several Colombian civil servants, who originally graduated as engineers, were sent mainly to the USA and came back with M.Sc. or Ph.D. degrees. The second one is that since its creations, the Geological Survey has undergone many restrictions and modifications, but it has remained as one of the most respected scientific institutions of the country.

Table 8.1 Creation of programs in geology in Colombia

Date	Name	Level	Institution	Locality
1942	IGP, IG	Undergr.	U. Nacional	Medellín
1956	GG, G	Undergr.	U. Nacional	Bogotá
1980	IG	Undergr.	UPTC	Sogamoso
1982	Geology	Undergr.	UIS	Bucaramanga
1983	IG	Undergr.	EIA	Medellín
1983	G	Undergr.	U. Caldas	Manizales
1983	Geology	Undergr.	E. EAFIT	Medellín
1985	Geophysics	M.Sc.	U. Nacional	Bogotá
1998	Geom./Soils	M.Sc.	U. Nacional	Medellín
1999	Earth Sc.	M.Sc.	U. EAFIT	Medellín
1999	Geology	Ph.D.	U. Nacional	Medellín
2010	Geology	Undergr.	U Pamplona	Cúcuta
2013	Geoscience	Undergr.	U. Andes	Bogotá
2013	Earth Sc.	Ph.D.	U. EAFIT	Medellín
2014	Geology	Undergr.	U del Norte	Barranquilla

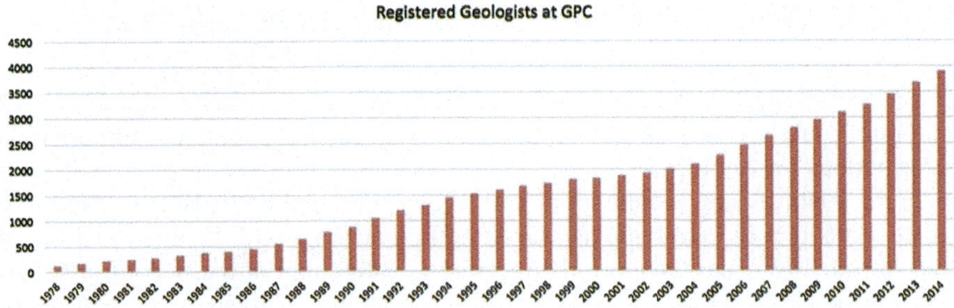

Fig. 8.1 Total number of registered geologists

8.4 Associations

The first organization was founded in 1957 under the auspices of several oil companies working in Colombia at that time: Sociedad Colombiana de Geólogos y Geofísicos del Petróleo (Colombian Society of Petroleum Geologists and Geophysicists, as an associate of AAPG). It is still active and has organized several national and international scientific meetings on petroleum geology.

Then followed the Sociedad Colombiana de Geología (Colombian Society of Geology) which was founded in 1960. Despite several ups and downs, the society has managed to survive and with the support of institutions as Servicio Geológico Colombiano, Consejo Profesional de Geología (National Council for Geologists), to organize on a regular basis the National Geological Congresses (the 15th will be in August 2015) and also international meetings in several Colombian cities. The society has also several regional chapters and has published several books on Colombian geology. An interesting activity in the last two decades has been the support to biannual meetings (Semana Técnica, the technical week organized by geology students of the different universities) in order to foster their managing abilities and their ethical values. Furthermore, the society publishes a weekly informative bulletin, Geonotas, which is distributed to all the geologists registered at the National Council, and monitors an electronic page with valuable information.

In 1971, the Colombian Society for Geotechnics started its activities and is still very active, particularly in Bogotá. Its members are mainly civil and geological engineers.

Later on, graduates from the Medellin School of Mines founded AGEMPET (Association of Geological, Mining and Petroleum Engineers) and those from Bogotá founded AGUNAL (Association of Geologists graduated from the National University). These two associations have not been very active.

8.5 The 09 Law of 1974

Since the beginning of the 1970 decade, a group of geologists was unsatisfied by the negative of the Engineering Professional Council to recognize Geology as a profession on the same level as engineering. Thanks to the efforts of a small group of geologists and geological engineers, Law 09 (1974) was approved by both national legislative bodies. Its main aims are:

- To create the Geology Professional Council.
- To require a professional registration (matrícula professional) in order to exercise Geology as a profession.
- To state the requirements in order to obtain the professional registration.
- To define duties for the Geology Professional Council.
- To specify professional activities which they must be carried on by geologists.
- To establish the prohibition to hire persons with no professional card.
- To fix the minimum percentage of Colombian geologists who should be contracted by private companies.

At the present time, geological engineers are required to obtain their professional card from the Engineering Professional Council.

The Geology Professional Council meets every month, attended by representative of each one of the universities with a program in geology. Representatives of Geological Engineering schools are permanent guesses. The CPG (the initials in Spanish) maintains close ties with the Colombian Geological Society and participates in the preparation of national scientific events and in the financing of publications of national interest (www.CPGColombia). It assumes the role of counselor for departments of Geology who require its advice. It keeps the profession register and publishes statistics. Within the national geological congresses, it organizes regularly forums on education and ethics and serves as a permanent contact among the schools of geology. Under the agreement signed with the Colombian Geological Society, it has helped to partially finance meetings and publications.

8.6 The Code of Ethics for Geologists

It was adopted by the CPG by Resolution 533 (1985). In its Chap. 1 (General principles) and 2 (duties and obligations of Geologist), it contains aspects which are common to the deontology of any profession related to engineering.

Until now, very few cases of faults committed against the Code have been received. Most are related to document forgery by people who never graduated as geologists and intent to figure as professionals in this field of knowledge. The CPG is presently preparing a complement of the Ethical Code for Geologists which will be presented for discussion to the Colombian geological community during next

National Geological Congress and will include the following topics (adapted from Peppoloni and Di Capua 2014):

- Geologists are aware of the social and ethical implications of their professional practice. They do their best for the geosphere protection for the present and future benefit of mankind within the concept of sustainable development.
- Geologists put the interest of society in the first place.
- Geologists never misuse their geological knowledge, even under constrain.
- Geologists are always ready to provide their professional expertise in case of urgent need.
- Geologists continue to improve their geological knowledge lifelong and maintain their intellectual honesty at work, being aware of the limits if their capabilities and possibilities.
- Geologists act to foster progress in geosciences, the dissemination of geological knowledge and the spreading of a geoethical approach to land planning and natural resource exploitation.

8.7 Further Considerations

Geologists trained in Colombian schools do not receive any formal course in Geoethics. One of the authors, who taught Environmental Geology during many years, remembers the surprise which generally assaulted students when they realized that mining or petroleum exploitations might have disastrous consequences for human health or nature preservation. On the other hand, their appreciation of global change and its relation to human behavior is poorly known and in many cases does not deserve special attention.

Furthermore, among a small group of geologists, environmental rules are seen as useless caprices imposed by "ecolatric" governments converted by biologists. This is of course an extreme position.

In Colombia, environmental concern started in 1974, when a law on natural resources and nature protection was approved by legislators. In the following decade, the occurrence of several natural disasters as earthquakes, floods, and volcanic eruption caused much havoc and left many victims. Governments reacted and produced laws to prevent and mitigate natural risks. The culmination of this tendency was the approval in 1998 of the law on land use planning, which requires every city, town, and village of the country to prepare such a plan with the participation of several professionals, including geologists. Besides this landmark, the administration of natural resources and risks is divided into many different entities depending on several ministries, a situation which makes their management and coordination quite difficult.

Are geologists duly prepared in order to assume these tasks? In principle, if they have taken a decent course in Environmental Geology, the answer could be affirmative. For the time being, it is out of question to introduce a new course as

"Geoethics" in the already overloaded undergraduate programs. But basic notions could intelligently be included within the Environmental Geology course.

In an Andean country like Colombia, the occurrence of multiple natural risks poses a heavy responsibility for young geologists who sometimes are faced with difficult decisions or at least recommendations. In this sense, a revision of the Environmental Geology courses would certainly be useful and may be recommended to the CPG.

Another aspect which is worth considering is the divulgation of Earth Sciences: This seems particularly important in a country where basic education programs do not contain a single course in Earth Sciences and where religious images are still taken in procession in order to conjure up volcanic eruptions! It is not only a matter of scientific culture; it may be quite often a question of survival. Who, if not geologists, should be responsible to make basics geological knowledge available for society?

A further practical point: it is important to convince students that participation in scientific and professional associations should be taken seriously. Many organizations in this country have seen their future in jeopardy because members do not pay their dues or because those elected as members of the board of directors do not even bother to attend the meetings. It should be stressed that these types of participation are an important contribution to the future of science and to a profession in particular and deserve sacrifice of spare time and discipline.

Finally, geologists should convince themselves that the preservation of the geological heritage of the country (fossils, landscapes, outstanding outcrops) is one of their important duties, which must be carried out through scientific and professional organization commitment.

References

Colmenares G (1973) Historia económica y social de Colombia, 1537–1719. Universidad del Valle, Cali

Hermelin, M. 1998. Actualización y modernización de la currícula en Ingeniería Geológica y Geología. In: Actualización y modernización del currículo en Ciencias de la Tierra en Colombia, Ed: ICFES, 1:14–34. ISBN: 958-680-028-8

Peppoloni S, Di Capua G (2014) The meaning of geoethics. In: Wyss M, Peppoloni S (eds) Geoethics, ethical challenges and case studies in Earth Science. Elsevier, Amsterdam, p 450

West RC (1972) La minería de aluvión en Colombia durante el periodo colonial. Universidad Nacional de Colombia, Bogotá

Chapter 9
The Geoethical Promise and the Incorporation of Disaster Risk Management in the Territorial Ordination of Colombia

Jorge Alberto Hernández-Restrepo

Abstract Colombia has been a pioneer in Latin America in its development of a comprehensive vision for the treatment of risks and disasters, allowing a reduction in loss of life, but damage to property, infrastructure and livelihoods continues to increase, which shows that disasters are not events generated by nature itself, but are the result of the application of inappropriate development models, which do not consider the society and nature relationship. Due the above, it is imminent to make transformations that incorporate restrictions and potentials according to the existing hazards, where the use of land, under the growing pressure for urban development, will be controlled in the future. Disaster Risk Management should be a social process, boosted by Geosciences, aimed to the formulation, implementation, monitoring and evaluation of strategies, plans, programmes, regulations and permanent actions for the risk identification and knowledge, as well as the possible forms for their mitigation, with the fundamental purpose of contributing to the security, the well-being, the quality of life of the people and the sustainable development. Geoethics stand as the fundamental discipline, to guide the decisions that must be implemented in order that the Disaster Risk Management can be incorporated in the territorial order, as the fundamental axis of development. That's why the work that must be performed by Earth Science professionals should be in consideration of Geoethics and its principles.

Keywords Geological risk in Colombia · Management of disasters

J. A. Hernández-Restrepo (✉)
IAGETH Working Group on Astrobioethics, International Association for Geoethics (IAGETH), Vancouver, NC, Colombia
e-mail: geo.jorge.hernandez@gmail.com

R. D. Acevedo and J. M. Frías (eds.), *Geoethics In Latin America*,
The Latin American Studies Book Series,
https://doi.org/10.1007/978-3-319-75373-7_9

9.1 Introduction

Due to its privileged location in the northwest corner of South America, Colombia offers innumerable advantages to live and enjoy its great potential of natural resources. It belongs to the equatorial zone of the planet and occupies space in the northern and southern hemispheres (Fig. 9.1). This privilege is framed by their coasts and territorial dominance in the Atlantic and Pacific oceans, three Andean mountain ranges with their Andean valleys, wide plains and rivers, besides being rich in jungles, forests and moors.

These benefits have brought in turn a number of difficulties that make the country susceptible of being affected by geodynamic events of endogenous and exogenous type, which in conjunction with extreme weather events, improper land use and conditions. Vulnerability of its population has generated diverse scenarios of risk by disasters. The Colombian territory is located in an area of very high tectonic complexity, where the Nazca, South America and Caribbean plates generate a high seismic and volcanic activity, evidenced by the occurrence of destructive earthquakes, tsunami and the recent activation of several of its volcanos. Just remember the tragedy caused by the Nevado del Ruiz volcano, which occurred

Fig. 9.1 Physical (blank) map of Colombia. Note both the Andean Ranges and the localization on the northwest corner of South America, which gives the country, territorial dominance over Atlantic and Pacific oceans. Modified from Shadowfox (2009)

Fig. 9.2 Aerial view of the buried town of Armero, after the volcanic eruption on 13 November 1985. 25,000 people were killed and 10,000 people were homeless. *Credit* Socialhizo.com (2016)

on 13 November 1985, when a Lahar caused by the melting glacier after an eruption, left buried more than 25,000 people in the towns of Armero–Tolima and Chinchina–Caldas (Figs. 9.2 and 9.3). In addition, the abruptness of its mountainous regions and the action of anthropic, biological and weathering agents such as rainfall, winds and temperature changes, characteristic of the climatic conditions of the tropics, have made Colombia a highly prone country to the action of severe events of erosion, landslides, torrential floods, floods and forest fires.

Colombia has been a pioneer in Latin America in its development of a comprehensive vision for the treatment of risks and disasters, allowing a reduction in loss of life. However, damage to property, infrastructure and livelihoods continues to increase, which shows that disasters are not events generated by nature itself, but are the result of the application of inappropriate development models, which do not consider the society and nature relationship. Despite efforts to contribute to the territorial security, social welfare and environmental sustainability, we must conclude that they have not been effective enough due to manifestations of conditions of increasing vulnerability. Hazards due to natural events, as well as global economic crises, climate change, environmental degradation, social inequality and armed conflict, are part of the spectrum of factors that must be considered in Colombia, so as not to jeopardize development of the nation.

Based on the above, it is imminent to make transformations that incorporate restrictions and potentials according to the existing hazards, where the use of land,

Fig. 9.3 A woman is rescued from the mud, some hours after volcano Nevado del Ruiz erupted. The eruption caused a massive Lahar, wiping Armero off the map. It was the first-known image of the tragedy and was called The Venus of mud. Felipe Caicedo (1985) on Tragedia de Armero: "Bajar la cámara y llorar". *Credit* Vargas (2014)

under pressure for urban growing will be controlled in the future, adjust technologies for interventions of ecosystems, among other aspects. Otherwise the economic losses and the effects on the population will continue to grow as they have done so far.

In this situational framework, Geoethics acquires a fundamental relevance. According to Rejas Ayuga et al. (2015) "Geoethics is the discipline that deals with an ethical approach, the professional and research subjects involved in Earth and Planetary Sciences as a whole". Its recognized importance at international level, as well as its aims and purposes, makes Geoethics stand as the fundamental discipline, to guide the decisions that must be implemented in order that the Disaster Risk Management, can be incorporated in the territorial order, as the fundamental axis of development.

Ideally, Disaster Risk Management should be a social process, boosted by Geosciences, aimed to the formulation, implementation, monitoring and evaluation of strategies, plans, programmes, regulations and permanent actions for the risk identification and knowledge, as well as the possible forms for their mitigation, with the fundamental purpose of contributing to the security, the well-being, the quality of life of the people and the sustainable development (Fig. 9.4). In Colombia's current legal system, this issue has become an indispensable policy, intimately associated with safe development planning and sustainable territorial environmental management, which must be supported and materialized at all levels of government, the Geoscientific community and private entities, with an effective participation of the people, who must seek their self-care.

In order to ensure sustainability, territorial security, collective rights and interests, improve the quality of life of communities at risk, it is fundamental that Geoscientists are a guarantee and a fundamental part of the adequate inclusion of

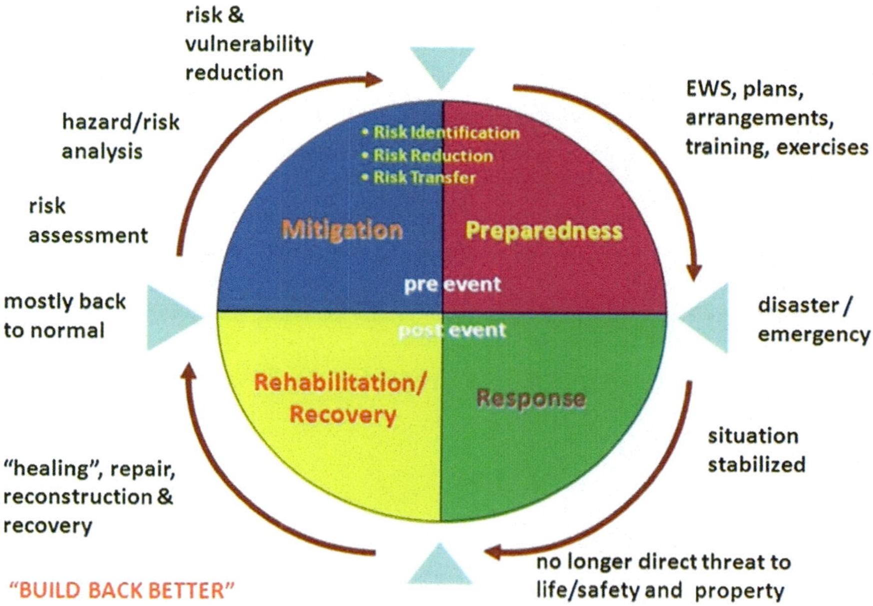

Fig. 9.4 Comprehensive Disaster Risk Management framework. *Credit* Alexina Watts (2012)

Risk Management, in the ordering of the territories, both at the municipal level as river basins. Given the above, it is important to think about the work that must be performed by Earth Science professionals in consideration of Geoethics and its principles.

Some members of the Italian Commission on Geoethics published an interesting article, on the Journal of the International Union of Geological Sciences (Matteucci et al. 2014), in which they proposed a Hippocratic-like oath for Geoscientists, called the "Geoethical Promise", which was slightly modified by the International Association for Promoting Geoethics (IAPG) at their Executive Council Meeting, hosted at the 35th International Geological Congress in Cape Town, South Africa, on 26 October 2016. This oath, built on the basis of the high ethical responsibility of Geoscientists, regarding the preservation of the planet, the proper use of the resources it provides us, as well as the protection of people and the achievement of sustainable development, will be used as a reference and reflection medium of the mission and the role that Earth Sciences play in Disaster Risk Management.

9.2 The Geoethical Promise

I promise…

…I will practice geosciences being fully aware of the societal implications, and I will do my best for the protection of the Earth system for the benefit of humankind…

...I understand my responsibilities towards society, future generations and the Earth for sustainable development...

...I will put the interest of society foremost in my work...

...I will never misuse my geological knowledge, not even under constraint...

...I will always be ready to provide my professional assistance when needed, and will be impartial in making my expertise available to decision makers...

...I will continue lifelong development of my geoscientific knowledge...

...I will always maintain intellectual honesty in my work, being aware of the limits of my competencies and skills...

...I will act to foster progress in the geosciences, the sharing of geoscientific knowledge, and the dissemination of the Geoethical approach...

...I will always be fully respectful of Earth processes in my work as a geoscientist.

I promise!

9.3 The Dissertation

- I will practice Geosciences being fully aware of the societal implications, and I will do my best for the protection of the Earth system for the benefit of humankind

Disaster Risk Management, as a scientific and social matter, is totally dependent on the Earth Sciences. The zoning of danger and risk by different kinds of natural events or those where humans in its interaction with the earth, cause disaster situations, should focus on the protection of life and property of those who inhabit each territory, avoiding the generation of new risk scenarios and environmental degradation.

Where there is a possibility of serious or irreversible damage to lives, property and rights of individuals, institutions and ecosystems, as a result of the risk situation becoming a disaster, the precautionary principle should be applied, whereby lack of full scientific certainty should not be an impediment to take measures to prevent or mitigate the risk (Fig. 9.5).

- I understand my responsibilities towards society, future generations and the Earth for sustainable development

The evolution of Geosciences has had a great impact on the development of humanity. In the last decades, this relevance has been increased by the application of Geoscientific knowledge, to improve the living conditions of people and the protection of their environment. Geoscientists must be fully aware that the application of his knowledge in Disaster Risk Management, in order to obtain the results more adjusted to the reality of the territory, is an undelegable compromise that affects the life of the communities and future generations.

Fig. 9.5 Precautionary principle versus burden of proof. *Credit* BioNinja.com (2017)

Development is sustainable when it meets the needs of the present without compromising the ability of environmental systems to meet future needs and involves taking into account the economic, social and environmental dimension of development. Disaster Risk is derived from processes of unsustainable land use and occupation. Therefore, rational exploitation of natural resources and environmental protection are irreducible sustainability characteristics and contribute to the Disaster Risk Management.

- I will put the interest of society foremost in my work

Although the primary purpose of research on Geosciences is the abiotic world, society is the final recipient of the results obtained through the research and actions carried out by the Geoscientist in Disaster Risk Management. For this reason, each hazard and risk zoning map, each project, each solution for Disaster Risk Mitigation, should focus on the protection of people and their assets. As stated by Jordan (2013) *"Geoscientists have a great responsibility towards society in the natural hazards management and their role is crucial to reduce the impact of natural phenomena and to improve the resilience of communities to future disasters"*. The great environmental challenges affecting human communities require not only a strictly scientific and technical preparation of Geoscientists, but also a reflection on their broader obligations towards society.

It is important that Geoscientists consider Geoethics as an indispensable framework for basing their training and activity. The principles of Geoethics can guide them to pursue the common good by weighing the benefits and costs of each election and identifying solutions that respect the environment and are friendly to society, ensuring respect for the right balance between human life and Planet Earth dynamics.

- I will never misuse my geological knowledge, not even under constraint

The phrase "*Scientia potentia est*" or "Knowledge is Power", is commonly attributed to Sir Francis Bacon (Bacon, Sir Francis, 1597). Knowledge is acquired in terms of seeking a truth or explaining a phenomenon. In a society, the function of those who define the truth is the transmission of this knowledge, which is done through norms and behaviours. Therefore, in a society exercising knowledge is synonymous with the exercise of power.

Advanced knowledge and skills that have been obtained by the Geoscience disciplines must applied be in favour of society and not against it. Due the above, this implies moral obligations, especially considering practical consequences, that are issues dealt with Geoethics.

- I will always be ready to provide my professional assistance when needed and will be impartial in making my expertise available to decision makers

Geoscientists have specific knowledge and skills, which are required to investigate, manage and intervene in various components of the Earth system to support human life and well-being, to defend people against geohazards and to ensure natural resources are managed and used sustainably. This entails ethical obligations. Therefore, Geoscientists must embrace ethical values in order to serve the people in a good way.

Geology is one of the main factors that most influence human development and determines the evolution of our society. It should be directed to respond to the challenges posed by the growth of urban areas in developing countries, since land management in these areas should directly influence the welfare of society.

> Geoscientists have new skills and tasks linking different disciplines, applying different methodological procedures and technologies, and facing new scientific, social and cultural challenges, from micro- to macro-scale studies, and from land, atmosphere and oceans to planetary exploration (Martínez-Frías et al. 2011).

- I will continue lifelong development of my Geoscientific knowledge

Geoscientists must be aware of their ongoing, voluntary and self-motivated pursuit of technical knowledge in order to acquire the skills they need to fulfil their aspirations and contribute to their societies. Geosciences have major impacts on the functioning and knowledge-based modern civilizations.

But, not only a day by day improvement of their professional skills are needed. Geoscientists have to improve their awareness of the Geoethical dimension and the importance of their work, for sustainable development and Disaster Risk Management. Only by increasing researchers' awareness of the ethical implications of their work, is possible to develop excellent Geoscience to serve and help society to be resilient and to reduce both the human impact on the environment as the configuration of new disaster scenarios, due wrong development policies.

- I will always maintain intellectual honesty in my work, being aware of the limits of my competencies and skills

The work of the Geoscientist in Disaster Risk Management implies a great responsibility towards the community of the area that is the subject of a specific study, since each result is used to make management decisions and establish control measures for new or existing human settlements in the territory, that can affect in good or bad way the lives and properties of the people. Geoethics should guide the professional activities, so that in their right measure they are carried out without overreaching, omission or overvaluation of propensity factors, intentionally.

At this point, it is important to emphasize that the methods of research and zoning of hazards and risks must be adjusted to the realities of the Colombian territory. In the country, disasters or their potential occurrence, most of them do not correspond to large events that cause a large volume and deployment of international aid, but rather, situations of low magnitude but high impact, due to the high vulnerability conditions of most of the population.

- I will act to foster progress in the Geosciences, the sharing of Geoscientific knowledge and the dissemination of the Geoethical approach

Communication and dissemination of the achievements of Geosciences must become key activities in building a knowledge-based society, able to better protect themselves and the Earth's ecosystems to ensure a life in harmony with our planet, for future generations. It is essential to keep all people informed, on Risk Possibilities, Disaster Management, Rehabilitation and Construction Actions, being respectful of the cultural particularities of each community and taking full advantage of the cultural resources of the same.

Geoscientists must strengthen their communication skills, overcome the communication barriers and think about how they can better ensure their knowledge and information is presented to the public (Fig. 9.6). But the question is not only how can the data and results can be presented, it is how the information can be showed well and effectively. If so, Geosciences can contribute significantly to a safer, healthier and more prosperous world.

- I will always be fully respectful of Earth processes in my work as a Geoscientist

With an age of more than 4.5 billion years, Planet Earth is the support of life, which has evolved in it for about 3.8 billion years. These beings, who have been successively populating the Planet, have taken advantage of it throughout this time. However, the human being, society in general, has little knowledge of the stage in which our life and progress have developed. It is little aware of the fragility of the planet and the need to respect and preserve it.

Knowledge of the Earth Systems (geosphere, atmosphere, hydrosphere, cryosphere and biosphere), their processes and interactions, should be an essential basis for an appropriate management towards sustainability. Studies in Geosciences provide much of that knowledge, in the field of Disaster Risk Management and the protection of the environment from anthropogenic action. This is especially relevant at the Risk Knowledge—Information and Reduction Stages (Fig. 9.7).

As can be seen, the application of the Geoethical Promise to Disaster Risk Management in Colombia is also applicable to the entire context of the Americas

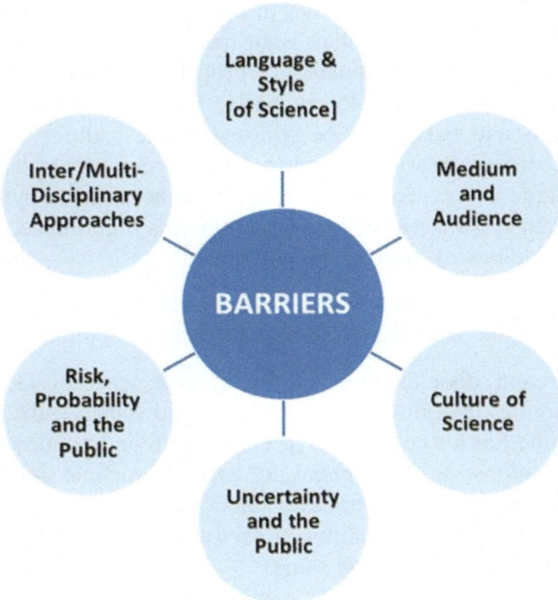

Fig. 9.6 Barriers to communication. Adapted from Beer (2010)

Fig. 9.7 How to increase Disaster Risk Management capacity? *Credit* The World Bank Group (2016)

and even the world. However, it is of great interest to note that part of the concepts and reflections included here are a fundamental part of the text and guiding principles of the so-called Disaster Risk Management Law in Colombia or Law 1523 of 2012, decreed by the Colombian Congress.

Geoethics must become an indispensable requirement of survival, mainly in Disaster Risk Management. The starting point that will allow us to constructively resolve the crises that affect us as beings interrelated with the environment is the recognition of the sacredness of all forms of life, the recognition of the interdependence between all living beings who are part of the biosphere and the recognition of the responsibility that corresponds to us to assume and to face the processes, events and phenomena that are manifested continuously in the planet as part of its natural evolutionary cycle.

References

Beer T (ed) (2010) Geophysical hazards: minimizing risk, maximizing awareness. Springer Science & Business Media. https://doi.org/10.1007/978-90-481-3236-2

BioNinja. Com (2017) Precautionary principle. Illustration Retrieved 15 Jan 2017, from http://ib.bioninja.com.au/standard-level/topic-4-ecology/44-climate-change/precautionary-principle.html

Colombia. Congreso de la República. Ley 1523 de 2012 (abril 24), por la cual se adopta la política nacional de gestión del riesgo de desastres y se establece el Sistema Nacional de Gestión del Riesgo de Desastres y se dictan otras disposiciones. Retrieved 12 Dec 2016, from http://www.secretariasenado.gov.co/senado/basedoc/ley_1523_2012.html

Jordan TH (2013) The value, protocols, and scientific ethics of earthquake forecasting. Geophys Res Abs 15(EGU2013-12789). EGU General Assembly 2013

Martínez-Frías J, González JL, Pérez FR (2011) Geoethics and deontology: from fundamentals to applications in planetary protection. Episodes 34(4):257–262

Matteucci R, Gosso G, Peppoloni S, Piacente S, Wasowski J (2014) The geoethical promise: a proposal. Episodes 37(3):190–191

Rejas Ayuga J, Bosque Sendra J, Antonio Malpica J, Maza Vázquez F, Dalda Mourón A, Soriano Sanz M, Rodríguez Díaz M, Luís Bermúdez J, Cerezal F, Goycoolea Prado R, González Matesanz F, Alonso Rodríguez M, Gómez Martínez F, Martínez-Frías J (2015) Aspectos geoéticos en la docencia del ordenamiento y la gestión del territorio. Ciencias Espaciales 8(1): 110–125. https://doi.org/10.5377/ce.v8i1.2044

Shadowfox (2009) Own work. Map. Retrieved 15 Jan 2017, from https://upload.wikimedia.org/wikipedia/commons/thumb/e/ee/Mapa_de_Colombia_(orograf%C3%ADa).svg/1377px-Mapa_de_Colombia_(orograf%C3%ADa).svg.png

Sir Francis Bacon, Meditationes Sacræ. De Hæresibus. (1597) Quotes (n.d.) Quotes.net. Retrieved 20 Dec 2016, from http://www.quotes.net/authors/Sir+Francis+Bacon%2C+Meditationes+Sacr%C3%A6.+De+H%C3%A6resibus.+%281597%29

Socialhizo.com (2016) Tragedia de Armero "Lo que el lodo se llevó". Photo. Retrieved 15 Jan 2017, from http://www.socialhizo.com/historia/historia-de-colombia/tragedia-de-armero-lo-que-el-lodo-se-llevo

Vargas R (2014) Tragedia de Armero: 'Bajar la cámara y llorar'. Photo. Retrieved 15 Jan 2017, from http://blogs.eltiempo.com/venga-le-cuento/2014/11/13/tragedia-de-armero-bajar-la-camara-y-llorar/
Watts A (2012) Addressing the risks & vulnerabilities of clients thru DRR-CCA A SEDFI sharing on DRR-CCA experience July 27, 2012. Diagram. Retrieved 15 Jan 2017, from http://slideplayer.com/slide/6237448/
The World Bank (2016) Disaster risk management in Armenia. Retrieved 15 Jan 2017, from http://www.worldbank.org/en/country/armenia/brief/disaster-risk-management-in-armenia

Part VI
Cuba

Chapter 10
Geodiversity, Heritage, and Geoethics in an Archipelago

Reinaldo Rojas Consuegra, Jorge Isaac Mengana, Rolando Cardenas, Kenya E. Núñez Cambra, Noel Pérez-Díaz and Osmel Martín

Abstract The Cuban archipelago has an extremely complex geology. In this chapter, an overview on its peculiarities is given. It is also presented a historical sketch of the Cuban geological heritage, and the current menaces to preserve it. Then, the current Cuban policy and practical efforts for geoconservation are described.

Keywords Geological heritage · Geopark · Chicxulub impact · Loma Capiro

10.1 Introduction

In several countries, geodiversity and geological heritage have been the focus of the geoscientific community for decades, whose recognition and management have reached the regulatory framework, for instance, the Natural Heritage and Biodiversity Law 42/2007 in Spain. In Cuba, during the first half of the 1990s, the need for a change in the man's approach to nature becomes clearer and more conscious in Cuban society. The anthropological egocentrism began to give way to ecological approaches, growing among different social actors, increasingly

R. Rojas Consuegra
Cuban Society of Geology, Havana, Cuba

J. Isaac Mengana
Humboldt Museum, Havana, Cuba

R. Cardenas (✉) · N. Pérez-Díaz · O. Martín
Planetary Science Laboratory, Universidad Central "Marta Abreu" de Las Villas, Santa Clara, Cuba
e-mail: rcardenas@uclv.edu.cu

K. E. Núñez Cambra
Institute of Geology and Paleontology, Havana, Cuba

R. D. Acevedo and J. M. Frías (eds.), *Geoethics In Latin America*, The Latin American Studies Book Series,
https://doi.org/10.1007/978-3-319-75373-7_10

educated and sensitive to changes in their natural environment. The legal regulations promulgated reflect the awareness of society in the struggle for a paradigm shift in the man–nature relationship, expressed in the Environmental Law 81 of the year 1997.

This decade was decisively marked by Cuba's own state policy. The economic crisis was intensifying and society was plunged to a critical situation, usually called as Special Period. Thus, the authorities took transcendental measures to guarantee the persistence of the most precious values of society. Among the first outcomes tourism, both domestic and foreign regained strength, and natural resources, particularly landscape and coastal areas, began to be used to raise foreign exchange for a resuscitative economy. The natural inheritance, and especially the patrimonial component in the island, including the geological one, had undergone above-mentioned avatars for more than four centuries. Today, there are various signs of growing awareness among various social actors. At this moment, the management policy of the geological heritage is evaluated as positive, a policy which highlights its patrimonial valuation, the value of use and the geoconservation projections.

10.2 The Cuban Archipelago: Peculiarities and Potentialities

The geodiversity of the Cuban archipelago is one of the highest in the world. Its formation occurred in a period of about 200 million years (Jurassic to present), in a complicated process that comprised sequences of rocks originating in at least three paleogeographic domains, namely the continental margin of North America, the continental block of Yucatán and the arcs of volcanic islands, elements evolved in space related to the opening of the Tethys Sea, the conformation of the Caribbean and its continental margins.

In this context, the petrological and lithological diversity is very outstanding, including a large variety of metamorphic, magmatic, and sedimentary rocks. In the same way, the fossil record is composed of more than a thousand species of fossils of both vertebrates, invertebrates, plants, and ichnofossils (Rojas Consuegra 2014). The complexity of the geological mosaic is also explicit in the diversity of soils, most of them formed under very dissimilar climatic conditions being distributed in a changing landscape, giving in some cases, a paradisiacal appearance.

A wide development of karstic conditions determines the existence of labyrinthine subsoil with large caverns, galleries, caves, furnias, dolinas, and valleys. In some cases, the superficial and subterranean waters form rivers and hundreds of seasonal currents; in other cases, make up significant aquifers of exquisite chemical compositions. At the same time, this dynamic and changing medium sustains several tropical plant formations in coasts, flatlands, and mountains, extremely rich in species and ecosystem diversity.

10.3 Geological Heritage

For decades, among the functions or interests of several Cuban institutions, the contribution to strengthening the creation of a scientific culture in society has been identified as a primordial task. In this sense, the socialization of geological knowledge of the country occupies a cardinal theme.

Nowadays, the Institute of Geology and Paleontology (IGP) represents the National Geological Survey (NGS). This institution has systematically contributed to geological research in the country, in collaboration with other institutions and universities. An important contribution to the geological research rests on post-graduate activities, in which master and doctorate programs play the major role.

A research supported by the IGP for years has been the inventory of the most outstanding geosites in the country. Thousands of localities have been assessed throughout the national territory, including the typical sections of the lithostratigraphic units that support the high degree of current geological mapping, deposits and mineral manifestations, sites and paleontological points, outcrops that contain particular geological contacts, folds, and lithologies, etc.

Also, the IGP has been promoting a broad and systematic work, from its updated corporate image, on the activities of the NGS: the printing of books such as guides and catalogs, brochures, and multimedia elaboration with the aim of impacting more effectively on a wide audience. Special mention should be made to the reopening of the institutional museum "Mario Sánchez Roig," an action that praises one of the most fecund researchers of Cuban Paleontology and Geosciences in general.

Another scientific–educational entity, the National Museum of Natural History of Cuba, an institution belonging to the Environmental Agency, has continued to provide postgraduate courses on Geosciences and their complex relationships with other disciplines for at least the last two decades. Topics such as geosciences and natural hazards, the paleoclimate and its validity before the climatic change, the fossil record with the teachings and heritage, the fossils and the architecture of the historic center of Havana among others have been considered.

An exclusive step has been marked by the summer courses on fossils, mainly aimed to motivate children and adolescents, although with open age participation (Rojas Consuegra 2013). Nevertheless, the main way and communicative support of this institution with its public are the exhibitions. Although modest, in them it is present the evolution of the organic world integrated to the own Earth´s evolution. An important number of Cuban fossils complement and support the teaching and museological discourse. Since 2009, the Museum maintains a Web page on the Paleontology of Cuba (www.redciencia.cu/webpaleo), where valuable information about fossils of the country can be obtained (Rojas Consuegra and Alabarreta Pérez 2009).

The work with scientific collections of rocks, minerals, and fossils constitute a reliable example of a mature curatorial research. Thousands of copies have been collected, cataloged, and identified in the last 50 years. Numerous fossil species have been made known for the first time not only in the country, but also for science, during the execution of numerous research projects (Rojas Consuegra 2015; Suárez et al. 2011). These treasures, an essential part of our scientific and cultural heritage, have been preserved despite the difficult economic conditions (Rojas Consuegra 2005).

Several investigations on the Cretaceous-Paleogene limit in the Cuban territory have been done. As a result, several areas and the scientific knowledge associated with this are now available for use in nature tourism (Rojas Consuegra and Núñez Cambra 2017). More recently, the Planetary Science Laboratory of Universidad Central "Marta Abreu" de Las Villas (UCLV) in Santa Clara (central Cuba) has engaged in research related with the Chicxulub asteroid impact, resulting in several international publications on the climatological and biological consequences of this impact (Pérez et al. 2013, 2014).

Another key institution preserving patrimonial value and geologic heritage in the archipelago is the Museum "La Casa Humboldt". Belonging to the Office of the Historian of Havana, this institution has maintained the education on paleontology with children of the community, by means of interests groups and workshops (Isaac Mengana et al. 2017). Also, it has used the architecture of the urban environment, rich in Cuban fossils, as didactic supports for teaching, and in addition, it has achieved a wide dissemination with plenty of visitors to the historic center of Havana (Isaac Mengana and Rojas Consuegra 2008, 2010). Also, excursions are done (Andares) in the city streets, with a high participation of the Cuban family (Isaac Mengana and Rojas Consuegra 2015).

Other scientific institutions (research and innovation institutes) and educational institutions (universities, polytechnics, and museums) also contribute with such work, provide specialized information on soil, water, landscapes, and other elements at different territorial scales.

10.4 Categorized Sites

Even when Cuba exhibits one of the highest geological mappings on a detailed scale (1:50,000) in the region, most of the geosites declared with some heritage category according to the current regulations of the country are caves or other karstic accidents, mainly due to the work carried out for long years by the Speleological Society of Cuba (SSC), founded in 1940 by Antonio Núñez Jiménez. It is worth remembering that on the occasion of the twentieth anniversary of this society, President Fidel Castro stated, pointing to the path of scientific development that has been projected for the country: Cuba will have to be, necessarily, a future of men of science.

Other remarkable paleontological sites have reached some patrimonial recognition as derivation of concomitant values, such as the fossil forest of Chorrillo (Sierra de Najasa, province of Camagüey) and other localities of fossil vertebrates, but largely because they are associated with karst.

The paleontological site "Domo Zaza" deserves special mention. This geosite was granted with the category of Outstanding Natural Element, based essentially on its paleontological attributes and values. Furthermore, this locality has given the only record of vertebrates' terrestrial fossils of the Neogene of Cuba, with exclusive taxa for the Science (MacPhee and Iturralde Vinent 1995).

10.5 Dangers and Threats

Unfortunately, throughout these five centuries of human occupation of the national territory, there have been losses of locations and materials due to multiple factors. The development of economic activities, such as construction of buildings, roads, and many others; intensive agriculture and changes in land use; the increase of the extractive industry of minerals and rocks among others, have represented immense pressure on geological heritage in general.

The transformation of karstic cavities for different purposes, such as the extraction of bat guano, the creation of storage spaces or crops, recreational facilities, has led to the loss of valuable and irrecoverable components of the diverse associations of fossil vertebrates that originated in them. The excavation and extraction of fossils, with an empirical character for hundreds of years, led to the formation of numerous collections of fossils containing thousands of pieces, but marked by a deep scientific and patrimonial bias.

At present, there are numerous localities distributed in extensive regions of carbonate rocks where it has existed since the days of the colony, the practice of sawing blocks of rocks of artisan form, for its use like material of construction. In this process, common fossils, mainly invertebrates, are generally revealed, which are generally not collected, so systematic loss of valuable scientific data is assumed. Also, there are extracted elements of vertebrates, mainly dental pieces of sharks and rays, sometimes of other fish, which usually go to private hands (Rojas Consuegra and Isaac Mengana 2008). In this case, it is not possible to identify the origin of the product.

The purchase of collections or pieces, in private hands, also seems to emerge today. This also raises interests for plundering and theft, including the state entities that are tenants of such real estate. Numerous pieces and significant collections brought to other countries in the past reveal the need to attend to their possible repatriations, or at least, take actions to know their contents, both quantitative and qualitative.

The opening of the country to a large international community, with the development of a growing commercial activity, in particular tourism, has also opened access to private collectors and amateur collectors. This situation is potentially detrimental to geological heritage in general, and in particular to paleontological heritage, taking into account the existence of traffic of patrimonial assets at international level.

On the other hand, alternatives as the collection of rescue in places in constant transformation, like many quarries, should be a practice to stimulate, above all, the scientific valuation of the materials.

10.6 Social Emergencies

A few years ago, local interests on the geological heritage, especially the paleontological heritage, began to be manifested, almost always combined with the archaeological or historical one. An interesting example is the community La Picadora (a cooperative of agricultural workers), in the municipality of Yaguajay, in the north of the province of Sancti Spíritus. It is the only one in Cuba that for more than a lustrum celebrates a scientific community event of Paleontology and Archeology, where dozens of specialists, including some foreigners, attend.

At the request of this community, a twentieth-century paleontological site has been re-studied. It is expected that the obtained scientific information contributes to the projection of a new museum where the paleontological–archaeological, and the historical, coexist in a harmonious way. Already some initiatives put into practice in this town bear fruit for the economic and social sustainability, where the geological and paleontological values of its surroundings are integrated, already with an identity character for its inhabitants.

In another case, a speleological and archaeological group has rediscovered a paleontological deposit, rich in fossil invertebrates, located near the town of Cruces, in the province of Cienfuegos. In addition to the new scientific results achieved, there has been a movement given in the interest of the community, where the provincial university plays an important role. Thus, courses have already been organized to involve different actors on the subject, and in other territories of the province, the dissemination of themes related to the local paleontological and archaeological heritage is achieved (Agüero Contreras et al. 2017). In particular, in the municipality of Rodas, the geological–paleontological, archaeological, and historical values are being recovered.

Interest in its research and geoconservation has been generated in localities where fingerprints of the K-Pg limit are detected, such as Loma Capiro near the city of Santa Clara, in the province of Villa Clara. The above-mentioned Planetary Science Laboratory (UCLV) in coordination with the National Museum of Natural History of Cuba and the Cuban Society of Geology (SCG) has been promoting the creation of a geosite, and perhaps later a geopark, at the Natural Park in where Loma Capiro is located. Similar efforts are underway in Fomento, in the neighboring province of Sancti Spiritus.

10.7 Facts and Projections

On March 1, 2014, the Council of Ministers of the Republic of Cuba approved the Geology Policy, whose main objective is "to generate and provide, in a systematic way, the geological knowledge of the national territory, with the purpose of preserving the geological heritage of the country."

In order to implement this policy, the Working Group for Geological Heritage was created to carry out a Geological Heritage Program, a process that included the participation of several agencies. As first essential aspects to deal with are a revision of the current regulatory framework and the proposal of the first geopark of the country: Viñales Valley, in the province of Pinar del Río. The Geological Department of the Ministry of Energy and Mines, among its strategic action program, has resulted in the creation of a group to follow everything related to this subject; in particular, it is focused on the constitution of the Geological Heritage Commission, attached to the National Heritage Center belonging to the Ministry of Culture, the rector organism of this activity in the country. The creation of the first geopark of the country in the world famous and beautiful Viñales Valley will complete the category that already bears this region as sociocultural landscape of humanity. With this action, the geodiversity and its georesources acquire a qualitatively new dimension for that exclusive environment, opening unlimited perspectives for the rest of the national territory.

The IGP-NGS develop new projects that will have a direct impact on the valuation of the country's physical environment, such as the completion of the Geological Map of Cuba to 1:50,000, and of the Geodiversity Map of Cuba, as a communication tool of the geoscientists with society. Other important actors contribute with these tasks, such as the geological enterprises and the universities, especially University Hermanos Saíz Montes de Oca and University Antonio Núñez Jiménez. Important subjects like the evaluation of geodiversity and the use value for different regions of the national territory are some of the topics considered in the projects. Also, the geological heritage has begun to be considered by disciplines like the social sciences (Center of Sociocultural Studies of the University Carlos Rafael Rodríguez).

The values of the geological heritage, and in particular the paleontological heritage, are now beginning to be seen, at least among a part of the Cuban geoscientific community, as a resource available for the sustaining of the current socioeconomic development (Rojas Consuegra 2015; Arano Ruíz et al. 2017). They also begin to transcend to other spheres, such as the social sciences, education, and culture (Rojas Consuegra et al. 2015; Agüero Contreras et al. 2017). In this sense, the Institute of Tropical Geography, belonging to the Environmental Agency (CITMA), works on updating the National Atlas of Cuba, where among others disciplines, geosciences will reflect their state of the art more updated.

Similarly, the Cuba Society of Geology maintains its quarterly newsletter. In addition, it sustains the annual delivery of numerous recognitions and prizes to the outstanding work of its affiliates. The association supports the improvement of its members and the initiatives in each of its territorial subsidiaries.

10.8 Natural Risks

The Ministry of Energy and Mines currently deals with the recovery of areas affected by nickel mining. Specialized companies deal with the ecological restitution of hundreds of hectares annually. These actions are complemented with broad participation and social communication, including the work with children of different levels of education.

Attention to natural hazards, vulnerabilities, and risks (HVR), along with industrial and public health, has become a political priority of the State and Government in Cuba. Virtually, all state agencies and bodies carry out actions for the prevention, mitigation, and eradication of HVR. The Civil Defense centrally attends the entire program of confrontation, where the preparation of the population plays a supreme role. In this context, the National Center of Seismological Investigations, belonging to the AMA (CITMA), maintains an active permanent monitoring of the earthquake-generating zones of the country, through an increasingly modern and comprehensive network of stations. Daily parts are issued with weekly summaries and other data of public use. The work is complemented with an efficient communication network that also favored the formation of an adequate perception of the risks in the population.

The continuous degradation of soils during several centuries of deforestation and aggressive agricultural practices had led to the deterioration of most of the soils of the island. Today, in response to the above situation, we consciously act to repair the soil conditions, a serious limitation to the sustainable development of food. Also, there has been an effective development of urban agriculture as a source of green and fresh products to the population of cities. Permaculture has been gaining adherents and expanding its benefits.

The Hydraulic Will was the policy on waters developed from the very beginning of the Cuban Revolution. Thousands of ponds and canals of different typologies have been constructed which collect alluvial and rainwater for its use in agriculture, livestock, industry, and supply to the population. These environmental transformations have brought ecological imbalances at local scale. Now, in correspondence with this situation, important studies are done to find feasible solutions with the aim of mitigating the impacts from a very general vision that included, among others, the protection of biodiversity, the shorelines, and the shallow bottoms of the broad marine platform. Also, this year is in approval the Water Law, a new legislation for the sustainable management of the vital liquid.

Other vulnerabilities related to climate change, such as coastal areas, hydrological basins, mountains, soils, and other natural components, impacting the population and the economy, are also promoted through different programs and projects in coordination with international organizations such as UNESCO, GEF, UNDP, WWF, and other donors. It is remarkable the recent call to scientists to contribute to the endeavor called "Tarea Vida," which is the government program to face the impact of climate change on the Cuban archipelago.

10.9 Geology and Health

Medical geology has ancient antecedents in Cuba, primarily in the traditional use of mineral-medicinal waters in many parts of the national territory, although only a few have reached our days with an almost uninterrupted supply of services (Elguea, San Diego de Los Baños). Sludge and peloids have been used in therapeutic practice in the country. At present, it would be desirable to continue the rescue of such georesources given their demand in nature tourism. Furthermore, in specific locations some researches have been done on the likely influence of the abiotic environment (rocks, soils, and water) on the genetic variation expressed in population health.

10.10 Protected Areas

In Cuba, the National System of Protected Areas (SNAP) has been systematically strengthened, having today an extensive network of protected areas, although its work has mainly focused on the values of biodiversity. At present, however, the SNAP also pays attention to the characterization of geological values in such areas with the identification of geosites (caverns and secondary formations, paleontological deposits, travertine formations, soils, etc.) which are protected according to the current geoconservation legislation. This step favors the use of such georesources making them accessible to different publics and particularly, to the current tourism.

10.11 Geological Tourism

Nature tourism is gaining new audiences in the country, especially, for foreign visitors, keeping in mind that the Cuban public has had for decades camping as a healthy option of rest and recreation. Usually, camping sites are inserted in places of exceptional landscape values. Regrettably, geotourism has not yet begun in our country in a widespread commercial way, but it is a practice that should have medium- and long-term projections considering the broad potential offered by Cuban geodiversity. This is an expressed interest of the community of geoscientists in the island, supported by a high geological knowledge available. According to the current policy on the diversification of tourism in the country, geotourism emerges as a very promising goal.

References

Agüero Contreras FC, Rodríguez Rodríguez RR, Borges Sellén CR, Arano Ruíz AF, Rojas Consuegra R (2017) Paleontología, Ciencias Sociales y Sociedad: una perspectiva para el desarrollo sustentable. En: Resumen XII Congreso de Geología, VII Convención de Ciencias de la Tierra. La Habana. Soc. Cub. Geología. CD ROM, Geo12-010. 1 p. ISSN 2307-499X

Arano Ruíz AF, Quintero Vázquez JS, Borges Sellén CR, Rodríguez Rodríguez RR, Rojas Consuegra R, Agüero Contreras FC (2017) Diversidad y potencialidad para la puesta en uso del registro fósil en la región de Potrerillo, Cienfuegos, Cuba central. En: XII Congreso de Geología, VII Convención de Ciencias de la Tierra. La Habana. Soc. Cub. Geología. CD ROM, Geo12-05. 1 p. ISSN 2307-499X

Isaac Mengana J, Rojas Consuegra R (2008) Patrimonio paleontológico en La Habana Vieja. Potencialidades educativas. En: 7mo. Congreso Internacional de Patrimonio Cultural: salvaguarda y gestión. 17 al 21 de marzo de 2008. La Habana

Isaac Mengana J, Rojas Consuegra R (2010) La Paleontología y los niños en el Museo Nacional de Historia Natural. En: Memoria del III Simposio de Museos y Salas de Historia Natural. 6 p. (CD-ROM). ISBN 78-959-300-009-3

Isaac Mengana J, Rojas Consuegra R (2015) Patrimonio paleontológico y valor de uso: una emergencia social. En: V Simposio de Museos de Historia Natural, X Convención Internacional sobre Medio Ambiente y Desarrollo. La Habana, 1 p

Isaac Mengana J, Rojas Consuegra R, Gutiérrez Gómez A (2017) Tesoros paleontológicos en la arquitectura de La Habana Vieja. En: VI Simposio de Museos y Salas de Historia Natural, XI Convención de Medio Ambiente y Desarrollo. La Habana. CD ROM, SMHN-40. 1 p. ISSN 2307-499X

MacPhee R, Iturralde Vinent M (1995) Origin of the Greater Antilles Land Mammal Fauna, 1: New Tertiary Fossils from Cuba and Puerto Rico. Am Mus Novit (3141):1–31

Pérez N, Cardenas R, Martin O, Rojas R (2013) Modelling the onset of photosynthesis after Chicxulub asteroid impact. Astrophys Space Sci 343:7–10

Pérez N, Martín O, Cardenas R (2014) Evolución del proceso de fotosíntesis después del impacto del asteroide de Chicxulub. Revista Cubana de Física 31:E54

Rojas Consuegra R (2005) Patrimonio paleontológico cubano: una categoría que merece ser reconocida y protegida. Sección técnica. Boletín de la SCG, Septiembre-Diciembre 5(3):14–15

Rojas Consuegra R (2013) Cursos de verano "Fósiles de Cuba" para niños y adolescentes en el Museo de Historia Natural: experiencia inédita. Boletín de la Sociedad Cubana de Geología 13 (2):12–13. ISSN 0864-3636

Rojas Consuegra R (2014) Columna ilustrada del registro macrofósil de Cuba. Revista Anuario de la Sociedad Cubana de Geología 2:13–18

Rojas Consuegra R (2015) Geodiversidad cubana: oportunidad de gestión del patrimonio geológico como un georrecurso para el desarrollo. En: V Simposio de Museos de Historia Natural, X Convención Internacional sobre Medio Ambiente y Desarrollo. La Habana. Memorias en CD ROM, 9 p

Rojas Consuegra R, Isaac Mengana J (2008) Desconocimiento del Patrimonio Paleontológico Cubano: una categoría emergente; su identidad y protección. Centr. Nac. Infor. Geol. (IGP). Geoinformativa 9:2–6

Rojas Consuegra R, Alabarreta Pérez N (2009) Sinopsis del registro fósil de Cuba. Sitio web sobre Paleontología de Cuba (2009–2017): www.redciencia.cu/webpaleo

Rojas Consuegra R, Núñez Cambra KE (2017) Guía para la excursión al límite K-Pg en Cuba occidental. En: Memorias de VII Convención de Ciencias de La Tierra. Sociedad Cubana de Geología. La Habana. Excursiones post convención No. 4. CD ROM, 34 p. ISSN 2307-499X

Suárez W, Rojas Consuegra R, Menéndez L, Silva G (2011) Informe final del proyecto "Biodiversidad paleontológica del archipiélago cubano: bases cartográficas y conservacionistas". PRCT-022. MNHN, AMA-CITMA. (CD ROM, inédito)

Part VII
Ecuador

Chapter 11
Ethics, Policy, and Risk Assessment of the Cotopaxi Volcanic Crisis in Ecuador—Vulnerable Society Versus Unprepared Volcanic Monitoring Staff and Authorities

Theofilos Toulkeridis, Ana Jacome and Fernando Mato

Abstract Risk monitoring, knowledge management, and communication skills play a key role in managing volcanic crises, and being the scientific knowledge, the transversal Achilles heel in risk perception, decision-making, general politics, and the main generator of confidence in population. During this study, we analyzed the ethical behavior of the policy in the risk management carried out in Ecuador on the Cotopaxi volcano after its most recent reactivation in 2015. This analysis clearly demonstrates evidence of both an inadequate risk assessment and an inadequate transmission of information to the authorities and afterward to the population, which led to loss of lives and a complicated social and economic situation throughout the affected zones. As a result of this study, important lessons were extracted and are provided here with the aim to prepare technicians and authorities to manage future volcanic crises in a more adequate form.

Keywords Ecuadorian ethics · Policy and risk assessment · Volcanic crisis
False alert · Inadequate planning · Cotopaxi volcano

11.1 Introduction

The responsibility of the evaluation of volcanic monitoring data, their transmission, and the corresponding communication from scientists to the authorities and afterward to the public awareness is fundamental (Newhall et al. 1999). In the last two centuries, volcanic hazards have claimed worldwide a few hundreds of thousands of

T. Toulkeridis (✉) · F. Mato
Universidad de Las Fuerzas Armadas ESPE, Sangolquí, Ecuador
e-mail: ttoulkeridis@espe.edu.ec

A. Jacome
Facultad Latinoamericana de Ciencias Sociales FLACSO, Quito, Ecuador

R. D. Acevedo and J. M. Frías (eds.), *Geoethics In Latin America*,
The Latin American Studies Book Series,
https://doi.org/10.1007/978-3-319-75373-7_11

people. A great part of such victims has been credited to wrong or false decisions in volcanic crisis situations especially during generation of lahars as shown in the case of the volcano Nevado del Ruiz in Colombia in 1985 (Sigurdsson, and Carey 1986; Naranjo et al. 1986; Tanguy et al. 1998). In the Andean, Cordillera of Ecuador is situated the Cotopaxi volcano, which is known for its far-reaching and lethal lahars generated due glacial ice melting during the last centuries (Barberi et al. 1995; Aguilera et al. 2004a, b; Pistolesi et al. 2014). As these lahars of the past threaten public and infrastructure alike, it is indispensible for the authorities to be correctly informed by the monitoring technicians and inform subsequently the public of its potential activity. The decision of who, how, and when to evacuate potential hazard zones threatened by lahars may be fundamental for the survival of villages and people living near the volcano (Haynes et al. 2008; Barclay et al. 2008).

Most importantly, the relationship between generation and clear interpretation of volcanic data during a reactivated volcano and the transmission of such information to the authorities and the responsible units and finally the subsequent decision-making and publication to the public decides about success in preventive activity or tragedy.

This case study will prove how a volcanic crisis went wrong as unprepared technicians were in command, when their false conclusions and wrong transmission of information were leading to panic with disastrous consequences. Furthermore, we will prove how unethical behavior took place, in order to cover traces of their misleading activities, of which consequences have led to the death of eight persons. The political background, which led to such situation and the role of the unprepared authorities will be also analyzed and presented. The aim of this study emphasizes that future volcanic catastrophes in Ecuador involving human lives are possible and most probably will take place because of the inadequate monitoring data handling and unprepared authorities.

11.2 Geological History and Volcanic Hazards Associated with Cotopaxi Volcano

The 5897 m a.s.l. high Cotopaxi stratovolcano is one of twenty active volcanoes, which are part of the northern cordilleran Andes in Ecuador being the result of the subduction of the oceanic Nazca plate below the South American and Caribbean continental plates (Freymuller et al. 1993; Toulkeridis 2013). The Cotopaxi volcano is situated between two major cities of Ecuador, being some 60 km south of Quito and 20 km north of Latacunga (Fig. 11.1). Two caters make up the top part of the current volcano, of which the more recently created crater is snow covered and is the site of volcanic vent activity. Two partial sector collapses occurred some 4600 and 2300 years ago forming debris avalanches and leaving some hummocks behind, while the usual volcanic activity is expressed by precipitation of ash and pumice as well as by pyroclastic flows, high gas emissions, lava flows, and lahars.

The volcanic history of Cotopaxi of the last four phases in 1534, 1742, 1768, and 1877 has been well documented by a variety of contemporary studies (La Condamine 1751; Sodiro 1877; Whymper 1892; Wolf 1878; Barberi et al. 1995; Aguilera et al. 2004a, b). Such studies described the past destruction of the close-by villages and surrounding infrastructure by a variety of up to 70 km/h fast lahars. Recent tephrastrigraphic and geochronological studies however evidence 19 strong eruptive phases with a re-occurrence of every 117 ± 70 years over the last 2200 years (Barberi et al. 1995). Furthermore, some additional 59 other minor volcanic events have been recorded between 1532 and 2015, 27 of which included lava flows and/or minor lahars besides regular gas emission and precipitation of pyroclastic material (Barberi et al. 1995). Therefore, statistic calculations based on the recorded events resulted in a probability of a new occurrence in the year 2015 of about 72%. Obviously, any re-occurrence of a minor volcanic event, like the 59 sporadic explosions, has been much higher.

Since the last severe volcanic event in 1877, a variety of strategic constructions and an enormous growth of villages and public have been established in the surrounding of glacier-capped volcano. While some 30,000 people lived in 1877 in the Southern, Eastern, and Northern hazard zone of Cotopaxi of which some 1000 persons lost their life, nowadays more than half a million persons habits the same danger zone. The impact time of the nearest city by a lahar is as low as 30 min from the formation of a lahar, with even less time for any warning for evacuation (Aguilera et al. 2004a, b).

11.3 Volcanic Risk Assessment in Ecuador

11.3.1 Authorities and Line of Command

Since 1988, the Instituto Geofísico of the Escuela Politécnica Nacional (IGEPN) is monitoring some of the active volcanoes in Ecuador. Since its foundation, this academic unit has been under the authority of the National Civil Defense (until 2008) and later below the technical (Secretaría Técnica de Gestión de Riesgo; 2008–09), national (Secretaría Nacional de Gestión de Riesgo; 2009–2013) secretary of risk management, today simply Secretaría de Gestión de Riesgo (SGR) with a rank of a ministry. Information on eruptions, and predictions of possible future eruptions, which are supported on direct observation and monitoring of the volcanoes in activity, is transferred after previous evaluation directly from IGEPN to the SGR. All volcanic alerts, as well as other alerts generated by a variety of natural hazards of either geologic or hydro-meteorologic origin, are emitted by the SGR, which directly informs authorities in charge of airports, water supplies, electricity, and security. The yellow, orange, or red alerts are spread later to the public via Twitter, e-mails, telefax, and other digital media by the executive secretary of the SGR.

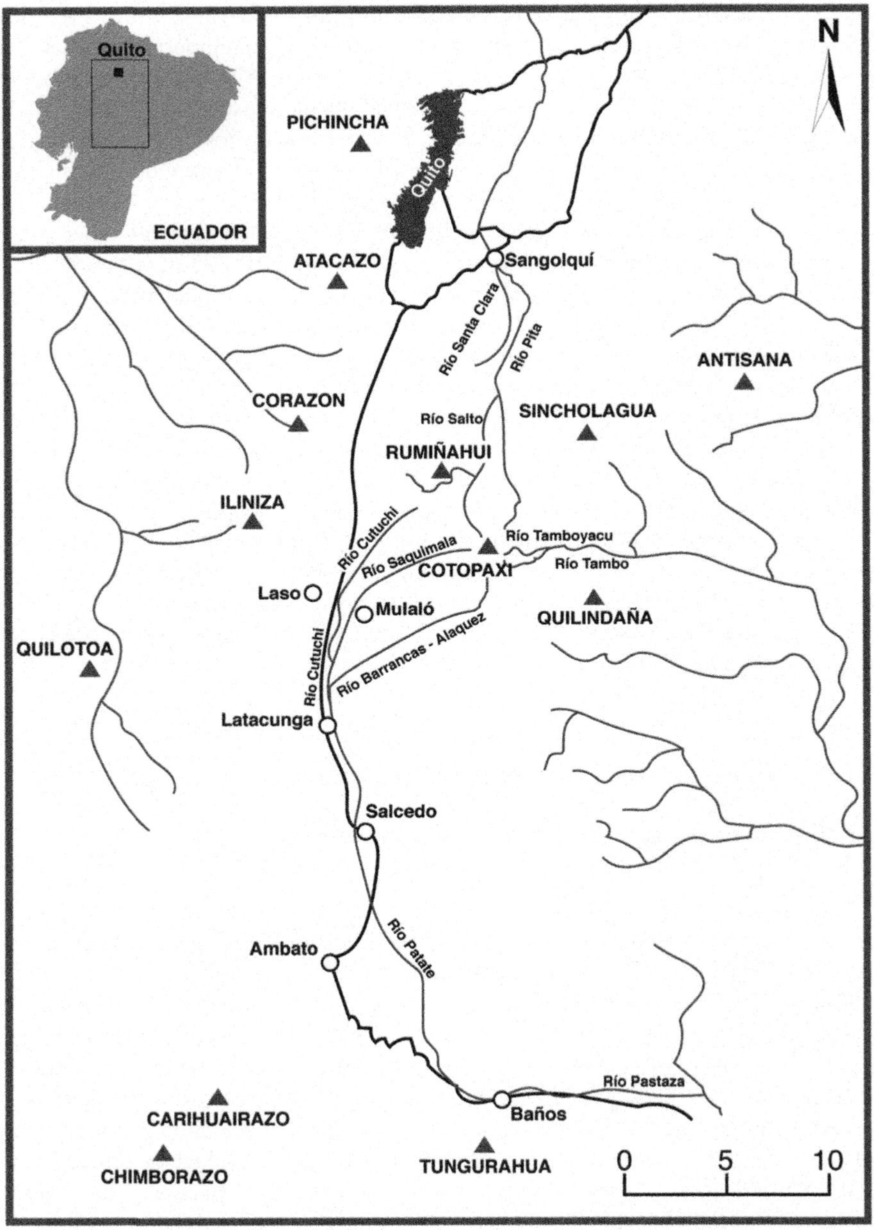

Fig. 11.1 Location of Cotopaxi volcano, rivers of lahar influence and flow and surrounding cities of a potential impact by future lahars (scale in kilometers). *Credit* T. Toulkeridis

11.3.2 Volcanic Monitoring in Ecuador

The permanent surveillance of the Ecuadorian volcanoes and the seismic activity in the country are the main function of the IGEPN, which is financially supported by the government, but also receives support from foreign agencies and research institutions. The IGEPN has state-of-the-art facilities as well as modern seismographs, pressure sensors, ground deformation, geochemical, thermal, and visual data (Kumagai et al. 2010). In particular, the IGEPN has installed at Cotopaxi volcano SP seismic stations and EDM control lines of which data are telemetered in near real-time via radio to Quito (Kumagai et al. 2010). There are also permanent seismic BB stations and infrasound sensors. These observations are supported by the incorporation of thermal and video imagery, SO_2 emission monitoring (COSPEC and DOAS), geochemical analyses, continuous GPS and tiltmeters, and micro-barometric surveillance (Arellano et al. 2008). Some of the work of the IGEPN based on field and instrumental data at Cotopaxi volcano has been published in recent years (Hall and Mothes 2008; Kumagai et al. 2010).

11.3.3 Education and Prevention

A few fundamental elements in prevention issues in the surrounding of Cotopaxi volcano should be presented in order to understand the level of preparation to face an upcoming volcanic crisis. These issues comprise an early alert system, evacuation maps, and drills as well as mitigation structures.

An early alert system independent of human decisions was proposed a decade ago (Aguilera et al. 2004a, b), but the IGEPN prevented its installation. Moreover, when this proposal became again public in June 2015 due to a series of conferences and a documentary, which explained how to use it (Toulkeridis 2005), authorities insisted to install another early alert system which they thought would represent an early warning toward the public concerning coming lahars. By the end of November, the instruments which were financially supported by the municipalities of Quito and Sangolqui included 16 seismic stations, five infrasound detectors, 13 lahar detectors, seven stations of video telemetry, one thermal camera, twelve deformation control lines, five inclinometers, five GPS stations, five stations of gas determination, one mobile multi-gas, ash-meters, and ten repeaters (El Telégrafo 2015). Unfortunately, what the authorities called an early alert system is an accumulation of different detectors, which together do neither warn the public in time nor are independent in the transmission of any early alert signal.

Maps of the Cotopaxi volcano and its lahars were presented for the first time by Wolf (1878) demonstrating the disaster of the same year. Later, further maps followed from the IGEPN in 1985 and 2003 scales of 1:50,000, while Aguilera et al. (2004a, b) published a map at a scale of 1:5000 in 2004. This map is the most precise and most used among the public as it was able to identify buildings and other infrastructure regarding the main drainage zones of the lahars. Nonetheless, as the IGEPN is in charge of volcanic monitoring, only maps of this institute were allowed to be called official. The municipalities used such maps and reproduced them to the public. However, the scale of the maps is too big (1:50,000) and introduces a big error when localizing the lahar risk zones. This was even noticed by the citizens and severely criticized. Errors included a vast extension of the lahar zones in populated and industrial areas where there was no occurrence of lahars in the past, rivers which "escaped" from the lahar flows, among other inconsistencies (Fig. 11.2). Even worse in such maps, the evacuation routes either indicated to enter from safe sites into the lahar zones or took into consideration much longer distances from a potential initial evacuation site into a safe zone (Fig. 11.2). Maps with a high resolution of both the lahar extensions and adequate evacuation routes were published by the Earth Science Department of the Universidad de las Fuerzas Armadas ESPE (Fig. 11.3), but such maps do not have the official support of the authorities.

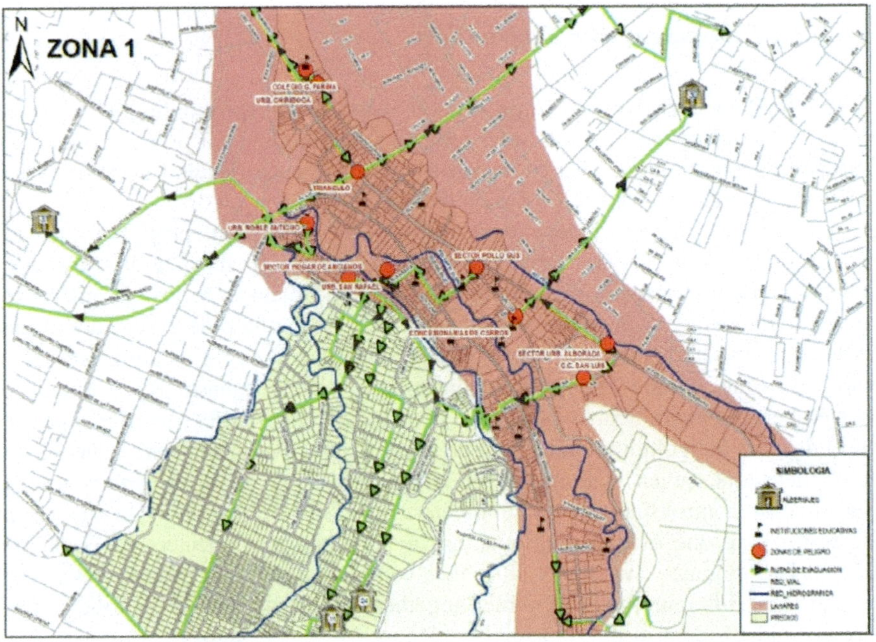

Fig. 11.2 Lahar and evacuation map of the Municipality of Rumiñahui of so-called Zone 1 (out of nine different continuing zones). Note that only main roads have been taking into consideration and often the recommended evacuation route is not the shortest. Also, evacuation directions are conflicting as demonstrated in the lower part of the map, where persons in a safe area are recommended to enter in the lahar area. All evacuation maps are lacking of a clear evacuation route policy. From Municipality of Rumiñahui (2015) *Credit* T. Toulkeridis

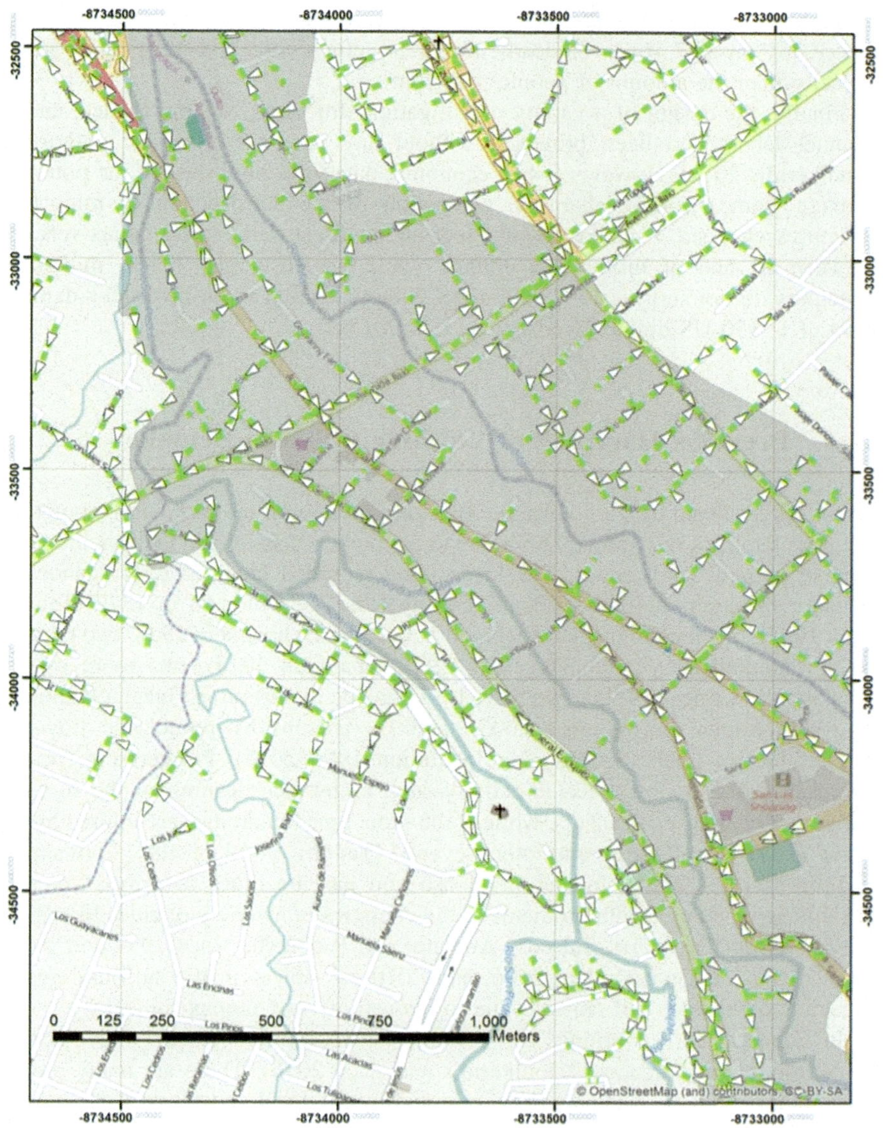

Fig. 11.3 Evacuation map of ESPE. Map is including the northern lahar zone in gray of the rivers Pita and Santa Clara, which end up in the river San Pedro. Taken from Padilla (2017) *Credit* Theofilos Toulkeridis

Evacuation exercises were partially executed in 2004 and 2005, as part of the reactivation of the volcano by the end of 2001 and repeated massively by the end of 2015. Nonetheless, due to the errors and inconsistencies in the official maps, the public remained mostly confused, as evacuation took longer that estimated leaded to "unsafe" zones or dead-ends or to refuge centers with a much lower capacity than the amount of people participating.

Finally, the design of a variety of mitigation structures, to stop, reduce, and/or channel lahars had been proposed without any response from the authorities (Toulkeridis 2010). However, when economic data were published of the potential damage from a future lahar, the perception of the construction of mitigation structures changed. Calculation of direct and indirect damage by lahars reached between 22 and 36 billion US dollars, while the costs of efficient mitigation structures do not surpass 150 million US dollars, leading to an invested—damage ratio of 1: 150 US dollars (Rodriguez et al. 2015).

11.4 Previous Handling of Volcanic Crisis in Ecuador

So far, direct fatalities from volcanic activities in Ecuador are rare but the potential of future fatalities is extremely high due to population density and their vicinity to a number of active volcanoes. The known and recorded fatalities by volcanoes in Ecuador reach some 1000 residents died by lahars of Cotopaxi in 1877 (Tanguy et al. 1998), one person by gas inhalation in Guagua Pichincha in 1948, two persons by strombolian activity in Sangay in 1977 (Snailman 1978), two more by gas inhalation in Chiles in 1990, two more by a phreatic explosion in Guagua Pichincha in 1993 (Annen and Wagner 2003), and six persons by pyroclastic flows in Tungurahua in 2006 (Smithsonian Institution 1999–2015). Furthermore, people deceased indirectly by volcanic activity such as by ash cleaning of the roofs in Quito in 1999 (2) and 2002 (1), while at the same time hundreds were injured by the same activity and these and other events (Tanguy et al. 1998; Smithsonian Institution 1999–2014). Inhalation of ash and gas are other described volcanic hazards aggravating health problems of the vicinities of a variety of cities like Quito (1999 and 2002), Baños, Salcedo, Ambato, and Riobamba since 1999 and in the coast mainly in Guayaquil in 2006, and 2010 as well as in the surroundings of Latacunga, Salcedo, and Ambato due to the activity of Cotopaxi in 2015.

In Ecuador, early warnings for volcanic events have repeatedly failed for all active-monitored volcanoes (Toulkeridis and Zach 2016). There are many prominent examples of such incidents, as reported by the local and worldwide press as well as the scientific community (Tobin and Whiteford 2002; Smithsonian Institution 1999–2015; Toulkeridis and Zach 2016).

The known record of failures starts with the declaration of yellow alert in early 1998 with the reactivation of the volcano Guagua Pichincha, which is situated on the western side of Ecuador's capital Quito. By the end of 1999, the orange alert was declared for about a week (from 27 September until 4 October), due to the growth and inflation of volcanic domes and the associated increasing number of seismic signals (Garcia-Aristizabal et al. 2007; Toulkeridis 2013). Nonetheless, although it is generally known, that volcanic domes tend to explode when seismic signals decrease during their "deflation," the alert status was downgraded to yellow during such phase. Due to this lack of knowledge, all following dome collapses with subsequent enormous explosions (VEI = 3) and ash precipitations (5 and 7 October, 26 November, and 10 December) occurred during yellow alert and not during a needed orange or red alert status (Garcia-Aristizabal et al. 2007; Toulkeridis 2013; Smithsonian Institution, 1999–2015).

A little later in the morning of the November 3, 2002, El Reventador volcano being the second most active in the country erupted with a VEI = 4, the strongest explosion in 140 years. No alert status at all was given until ash was precipitating 5.5 h later in Quito (Reischmann et al. 2003; Hall et al. 2008; Smithsonian Institution 1999–2015). In the following years, between 2002 and 2016, further volcanic activity occurred also without the necessary alert status given without any previous changes of criteria of the corresponding monitoring staff of the IGEPN and the corresponding confiding authorities.

One of the best examples of the mentioned wrong handling in risk assessment and early warning represents the Tungurahua volcano, which is situated on the eastern volcanic cordillera nearby the city of Baños de Agua Santa. This volcano re-awakened visibly by the end of 1999 and has been active since. Most eruptions of Tungurahua volcano appeared "suddenly"; therefore, the alert status, yellow, orange, or even red does not correspond with the current volcanic activity. The declaration of the yellow and later orange alert on the October 16, 1999, has been declared based on the publication of a picture with incandescent rocks on the top of the volcano and not based on any measurements of the installed monitoring instruments. As result of this declaration of the alert status, the city of Baños and nearby villages were completely evacuated for about three months (Tobin and Whiteford 2002). However, in the time of evacuation, the city of Baños has not been affected at all by the volcanic activity, including ash precipitations, as the main directions of the predominant winds have been away from the city (Toulkeridis and Zach 2016). Seeing from some distance such fortunate situation, in early 2000, the citizens of Baños and the nearby villages returned violently back to their homes ignoring further failed explanations and predictions of authorities and the IGEPN. Since this incident, the volcano alert status has been declared mostly to be orange without accomplishing any of the predictions set up by the monitoring unit of the IGEPN.

Later in 2006, a three months emergency declaration given by the authorities and the monitoring unit for the volcano Tungurahua just expired when the very first pyroclastic flows took place on the western side of the volcano (July 14, 2006) injuring some dozens of persons. In the following weeks, the volcanic unrest continued with regular appearances of a high amount of minor pyroclastic flows and lahars, when suddenly and without any specific warning a VEI = 3 eruption with a high amount of far-reaching pyroclastic flows devastated five villages and killed six persons in the early morning of the 17 August. The associated ash precipitation caused the closure of a variety of airports including Manta and Guayaquil. However, no airport has been warned for about eight hours after the eruption took place causing the return or deviation of several flights from Quito to Guayaquil, while flying through the ash clouds endangered the life of passengers and flying crew. Further upgrades of the alert status of most of the main eruptions of Tungurahua volcano after 2006 were declared two to four hours after the eruptions initiated. Therefore, in the following years, airports have been warned just shortly before (and often during or after) ash arrival (Toulkeridis et al. 2007; Smithsonian Institution 1999–2015; Toulkeridis and Zach 2016).

Demonstrating the lack of precise or even close warnings of volcanic eruptions and associated ash precipitations or other volcanic hazards of the few volcanoes which have been active during the last 16 years between 1999 and 2015 led to the concern of what else could happen if the other currently dormant volcanoes awake, like for example Cotopaxi volcano (Barberi et al. 1995; Biass and Bonadonna 2013).

11.5 The Recent Volcanic Activity of the Cotopaxi Volcano

11.5.1 Volcanic Activity in 2001–2015

After some phreatic explosions in 1940–42 and seismic and gas activity in 1976–77, the Cotopaxi volcano re-awakened by the end of 2001, showing some increasing seismic activity, gas emissions from a variety of fumaroles and the main crater, and also several small phreatic explosions (Cerca et al. 2005; Toulkeridis 2006; 2010). After more than a decade of minor activities, the seismic activity increased dramatically having a peak activity since April 2015. Seismic signs reached up to a few hundreds per day, while emissions of SO_2 reached up to 5000 ton a day (IGEPN 2015a, b). Therefore, a soon reactivation of Cotopaxi volcano seemed to be just a matter of days or weeks.

11.5.2 August 14, 2015—First Explosions

At 17:27 in the night of the 13 August, a clear precursor of an imminent explosion took place with a seismic activity being much higher than any registered in months. Just a few hours later, at 4:02 and 4:07 a.m. of the 14 August, the very first two explosions occurred, followed by three more at 10:25 a.m., 13:45 and 14:29 p.m. The technical and monitoring staffs of the IGEPN were not able to recognize the very first explosions, and it took a few hours until an official statement was send to the SGR and the public (IGEPN 2015a, b). The reason of such late statement was not the time it took to interpret seismic or other technical data, rather than that a high amount of citizens reported ash particles over their cars or other belongings and also that climbers reported by phone to have heard explosions in the early morning hours. The very first explosion at four in the morning has been the most voluminous and dense of the whole reactivation of the volcano in the time period of August until November. Nonetheless, it took some seven hours to declare yellow alert by the SGR and the IGEPN, as the first pictures taken by dozens of citizens and the press circulated by the social media of the visible eruption of 10:25 a.m. From this, as well as the afternoon eruptions, ash emissions precipitated mainly in the northern and eastern side of the volcano in the following hours. People remained calm and took the eruptive event as it has been a simple spectacle of nature keeping their respect and corresponding fear of what else can happen related to the volcanic activity of this volcano.

11.5.3 August 15, 2015—False Alert, First Victims, and State of Exception

As most of Ecuadorians have never witnessed before in their life volcanic activity of any kind coming from the Cotopaxi volcano, when the most feared national volcano showed signs of eruptive activity, wrong or inadequate information could lead to panic and chaos. At the 15 August, just one day after the reactivation of Cotopaxi volcano, the IGEPN informed the public at 7:40 a.m., the most feared message about the emplacement of pyroclastic flows and the generation of lahars on the southern flank of the volcano. The news have been spread first by the regular Twitter and Facebook sites to the thousands of followers and users of such social media but also via fax and phones to the authorities. These messages were supported by a picture, taken from the official monitoring cameras, apparently demonstrating the pyroclastic flow (Fig. 11.4a). As result of these messages, chaos and panic led to the uncontrolled behavior of the public and also of the authorities in Lasso, Latacunga, and other minor cities and villages at the southern surrounding area of Cotopaxi volcano. People afraid by the apparently arriving lahars were additionally warned by all kinds of news channels by the direct authorities like mayors, governor, and state prefect. This situation led police patrolling with high-speed ways and streets where people were also appealed to leave the area immediately. Such situation is rarely controlled, and therefore, it led to some seven mortal victims mainly in Latacunga, most by traffic accidents by run over and victims with sensitive health issues by heart attack or on the transport to far away hospitals (Ambato). As a result of numerical modeling (Barberi et al. 1995; Aguilera et al. 2004a, b; Pistolesi 2008; Pistolesi et al. 2014), it is known that lahars would impact Latacunga in 90 min after being generated. Nonetheless, the official message of the SGR to evacuate preventively was spread by Twitter at 9:00, meaning some 80 min after the initial messages of the IGEPN and 10 min prior a potential impact of a lahar in Latacunga. Also, a news bulletin with the description of the generation of pyroclastic flows and lahars was send to the press media.

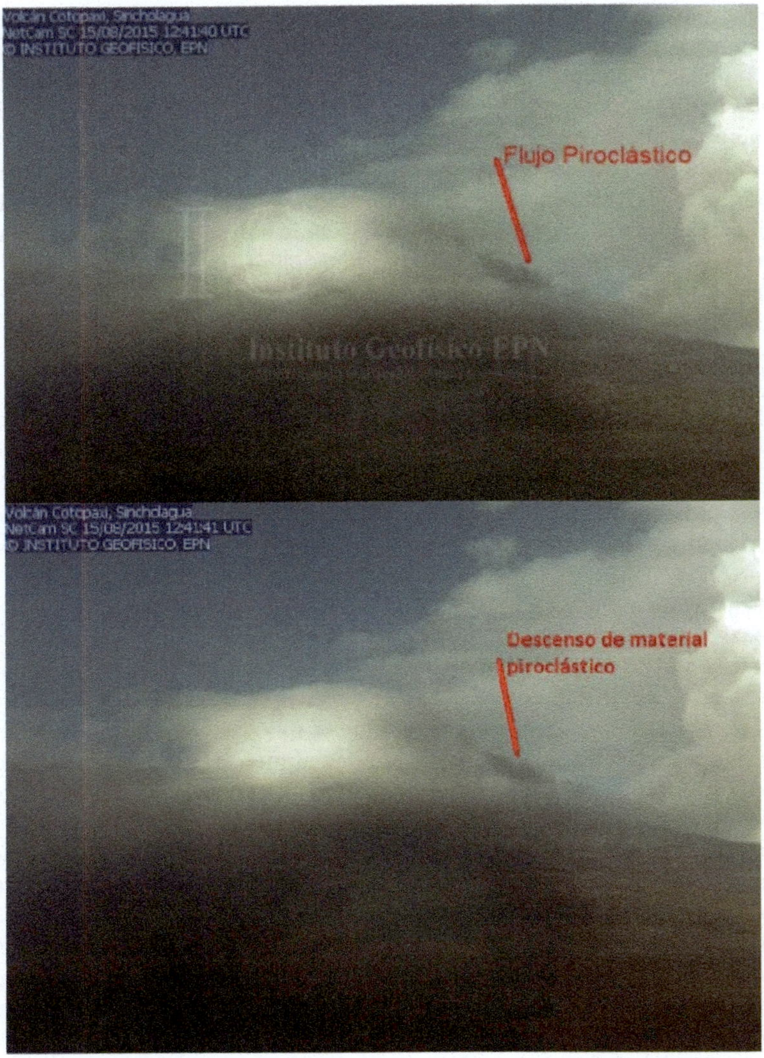

Fig. 11.4 **a** (above): Announcement of a pyroclastic flow (flujo piroclático) in the main Web site and twitter of the IGEPN (now deleted) of the morning of the 15 August. **b** (below) Same picture but with a different explanation—Fall of pyroclastic material (Descenso de material piroclástico) *Credit* IGEPN (2015c)

However, no lahar was seen by anybody for many hours. In fact, there has been no lahar at all during the activity of the Cotopaxi volcano in 2015.

Nonetheless, in the same afternoon, some nine hours after the initial warning of pyroclastic flow generation, which shall have produced lahars, a flight over the volcano reported that there has been neither any pyroclastic flow nor any generation of lahars. Apparently, it has been confirmed what seemed clear for many hours, that it had to be a false alert. The observations, the transmitted data of the monitoring as well as the subsequent decisions by the IGEPN about such volcanic hazards were absolutely wrong. An official statement via another news bulletin of the SGR followed. Additionally, the original photograph marking the apparent pyroclastic flow was changed into “fall of pyroclastic material” (IGEPN 2015c; Fig. 11.4b). All records in the official Web sites of the IGEPN about the generation of pyroclastic flows and lahars were deleted clearly proving unethical behavior in order to cover the responsibilities of the previously emitted warnings. Even more, no such report of what really happened in the morning of the 15 August has been send to the Global Volcanism Program of the Smithsonian Institution (Smithsonian Institution 2015). There it has been stated in the weekly reports “During 15–16 August sulfur dioxide emissions were high, and remobilized ash from the W flank rose up to 3.3 km; no ashfall was reported and only minor amounts of a sulfur odor were noted by residents.” which is a clear contradiction to the real circumstances of that day (Smithsonian Institution 2015).

Based on this clear failure in evaluation and communication of the real behavior of the volcano and the subsequent potential mistrust to the authorities in the handling of this situation, together with the social and political climate in Ecuador led the president of Ecuador to present and sign publically in the same afternoon the State of Exception for the whole territory of the country (Ecuadorian Presidency 2015).

11.6 Social and Economic Situation After False Alert: Knowledge Management and Communications Skills

During the time period of the State of Exception, news of any kind about the volcanic behavior was censored and those who spoke or wrote in or to the public about any volcanic issue were either firstly warned with imprisonment or called to explain their acting to the authorities (Pierson et al. 2014). No interviews were allowed in which one may express his or her opinion about the state of the volcano. Even universities and their scientists were warned to publish any scientific or technical results concerning the Cotopaxi volcano.

11.6.1 Knowledge Management

The state of exception is a powerful tool that facilitates the management of extraordinary situations for the national security. Due to the transversal implications of volcanic crises on multiple contexts (human, economic, social, psychological, political, etc.,), the state of exception allows a better implementation of the different states of alert in such disasters. This powerful tool, however, should not hinder the flow of information between the scientific community, decision-makers, and the public, as derived from The Sendai Framework for Disaster Risk Reduction 2015–2030 (United Nations 2015).

Firstly, scientific knowledge plays a key role in managing volcanic crises, being the Achilles heel in risk perception, decision-making, and the main generator of confidence on population (Marti 2015). It is for this reason that the interdisciplinary discussion and the scientific consensus become key to avoid errors that may lead in catastrophic situations, helping to provide on time, true, accurate, and clear information for effective decision-making. In this respect, Sendai Framework determines for academic and scientific networks as priority functions the following: (1) factors and risk situations; (2) applied research at local, national, and regional level; (3) support for communities and local authorities; and (4) science-State collaboration in decision-making. It is therefore necessary that the State consider the participation of academia and scientific networks in managing volcanic crises. Beyond restrict the participation, Sendai Framework emphasizes the role of all public institutions and society in disaster risk management, pointing the State as responsible of its prevention and reduction. The expected goal with the agreements reached at Sendai Framework is none other than the risk reduction, avoiding the emergence of new risks, and improving resilience. It is due to this reason that at this new framework "disaster management" was replaced by "disaster risk management," making reflect on the need for the Ecuadorian government to prioritize mitigation plans to avoid future disasters mainly caused by lahars.

Secondly, the Sendai Framework points to the state as head of the disaster risk management, determining the need to conduct a public report of such management. Regarding this, Ecuador should encourage the appropriate intergovernmental dialogue between authorities in the preparation of action plans, taking steps to prevent future lack of coordination (La Hora 04/08/2015) and ineffectiveness (El Comercio 08/09/2015) in decision-making.

11.6.2 Communication Skills

Equally important is communication skills in natural hazards, specifically in volcanic crises (Atman et al. 1994; Newhall et al. 1999; Visschers et al. 2009; Aspinall 2010; Donovan et al. 2012a, b; Stein and Geller 2012; Doyle et al. 2014; Marti 2015). This begins with a clear and on-time transmission to decision-makers about the diagnosis and forecasts derived from monitoring activities (scientists–decision-makers communication skills). At this level of communication, truthfulness and accuracy of the information transmitted is determined primarily by the quality of early warning system available, absent in the case of Ecuador. Secondly, for the quality of both the scientific data interpretation and the message to be transmitted. These reach a higher level and a lower time of availability as more heterogeneous is the range of experts involved in the process (scientists-to-scientists communication skills), restricted however in Ecuador to technicians (and a limited number of scientists) of the IGEPN. Academic discussion and broad scientific consensus, nonetheless, require a greater number of experts involved to avoid dramatic consequences of decision-making mistakes during disasters risk management, as was evidenced in Ecuador during different volcanic crises. Its major weakness is in the proper management of uncertainty (aleatory and epistemic) (Cox 2012; Marti 2015), critical point in Ecuador due to the absence of an early warning system and for the lack of information from IGPN to decision-makers, media, and population about uncertainty of forecasts. This fact has been aggravated by ambiguous reports limited to collecting the results observed through the monitoring process, significantly hindering the assimilation of information by decision-makers. The population, meanwhile, has been invaded by uncertainty and confusion, for the brief information derived from the following level of communication (decision-maker-population) devoted to: (1) formalize the transfer of scientific information to the public; and (2) inform, if applicable, about a possible warning change and the derived actions to be carried.

In an effort to avoid uncertainty, confusion and chaos during the crisis of the Cotopaxi volcano, the state of exception tried to achieve a controlled situation of information diffusion to the population. Sadly, the opposite effect was reached, mainly by the following reasons: (1) by the absence of a needed scientists-to-population communication (Marti 2015), not getting to correct however the appearance of disinformation news provided by sensationalist and lacking in rigor media, and the appearance of false rumors in social networks; (2) by the waste of a potential correct use of these social networks for handling this type of crises (Hiltz et al. 2011; (3) by the lack of coordination in decision-making (La Hora 04/08/2015; El Comercio 08/09/2015); (4) by the disinformation detected by experts and users inside improvised warning applications developed for mobile phones, implemented without the necessary supervision by experts and the SGR.

11.7 Conclusions and Recommendations

Firstly, it appears that the monitoring system of Ecuadorian volcanoes needs a fundamental improvement or even a radical change in order to develop a professional real-time data determination and a better knowledge of the responsibility of the way, how valid and clear information may be transmitted to the SGR, authorities, and public. To efficiently solve these gaps, firstly the implementation of an integral early warning system, with capabilities to integrate effective monitoring, accurate diagnosis, forecast, decision-making and skilled communication of volcanic activity in the country is needed.

Secondly, the following set of coordinated actions that provide a more scientific knowledge, expertise, know-how and professional ethics must be taken into consideration: (1) the implementation of a national research network, allowing sharing resources and scientific and technical cooperation with other regional and international networks, for research and for design mitigation plans in the medium and long term; (2) the creation of a scientific commission for the management of future volcanic crises (communication improvement in the scientists-to-scientists link), and for support political authorities in decision-making (communication improvement in the scientists-to-decision-makers link); (3) the implementation of a master plan for decision-making with a complete redesign of the interdisciplinary chain of command (communication improvement in the scientists-decision-maker link and in the decision-maker level); and (4) the implementation of an integral communication plan, which optimizes the information flow, its quality and its clarity (communication improvement in scientists-to-media and scientists-to-population links). This plan must provide proper tools at the same time for managing and sharing information between social networks and warning systems.

References

Aguilera E, Pareschi MT, Rosi M, Zanchetta G (2004a) Risk from lahars in the northern valleys of Cotopaxi Volcano (Ecuador). Nat Hazards 33(2):161–189

Aguilera E, Yépez W, Toulkeridis T (2004b) Early alert system for lahars of the volcano Cotopaxi (Ecuador). In: IAVCEI general assembly 2004, Pucón, Chile

Annen C, Wagner JJ (2003) The impact of volcanic eruptions during the 1990s. Nat Hazard Rev 4:169–175

Arellano S, Hall M, Samaniego P, LePennec J-L, Ruiz A, Molina I, Yepes H (2008) Degassing patterns of Tungurahua volcano (Ecuador) during the1999–2006 eruptive period, inferred from remote spectroscopic measurements of SO_2 emissions. J Volcanol Geotherm Res 176:151–162

Aspinall W (2010) A route to more tractable expert advice. Nature 463:294–295

Atman CJ, Bostrom A, Fischhoff B, Morgan MG (1994) Designing risk communications: completing and correcting mental models of hazardous processes, Part I. Risk Anal 14:779–788

Barberi F, Coltelli M, Frullani A, Rosi M, Almeida E (1995) Chronology and dispersal characteristics of recently (last 5000 years) erupted tephra of Cotopaxi (Ecuador): implications for long-term eruptive forecasting. J Volc Geothermal Res 69(3):217–239

Barclay J, Haynes K, Mitchell T, Solana C, Teeuw R, Darnell A, Crosweller HS, Cole P, Pyle D, Lowe C, Fearnley C (2008) Framing volcanic risk communication within disaster risk reduction: finding ways for the social and physical sciences to work together. Geol Soc London, Spec Publ 305(1):163–177

Biass S, Bonadonna C (2013) A fast GIS-based risk assessment for tephra fallout: the example of Cotopaxi volcano, Ecuador-Part I: probabilistic hazard assessment. Nat Hazards 65:477–495

Cerca M, Toulkeridis T, Concha-Dimas A (2005) First results about the structural setting and emplacement model of the Cotopaxi Volcanic Complex, Ecuador. In: Chapman conference on the effects of basement, structure, and stratigraphic heritages on Volcano behaviour, Taal Volcano, Tagaytay City, Philippines, 36

Cox LAT (2012) Confronting deep uncertainties in risk analysis. Risk Anal 32:1607–1629

Donovan A, Oppenheimer C, Bravo M (2012a) Social studies of volcanology: knowledge generation and expert advice on active volcanoes. Bull Volc 74:677–689

Donovan A, Oppenheimer C, Bravo M (2012b) Science at the policy interface: volcano-monitoring technologies and volcanic hazard management. Bull Volc 74:1005–1022

Doyle EE, McClure J, Johnston DM, Paton D (2014) Communicating likelihoods and probabilities in forecasts of volcanic eruptions. J Volc Geothermal Res 272:1–15

Ecuadorian Presidency (2015) Presidente de la República firmó el Decreto 755 (DECRETO). http://www.presidencia.gob.ec/presidente-de-la-republica-firmo-el-decreto-755-decreto/

El Comercio 08/09/2015 (2015) Las autoridades de Cotopaxi tardaron 4 horas en el simulacro de cambio de alerta de amarilla a naranja. http://www.elcomercio.com/actualidad/cuatro-horas-simulacion-autoridades-cotopaxi.html

El Telégrafo 10/12/2015 (2015) Parque Nacional Cotopaxi se reabre para visitantes. http://www.telegrafo.com.ec/regionales/regional-centro/item/parque-nacional-cotopaxi-se-reabre-para-visitantes.html

Freymuller JT, Kellogg JN, Vega V (1993) Plate motions in the north Andean region. J Geophys Res 98:21853–21863

Garcia-Aristizabal A, Kumagai H, Samaniego P, Mothes P, Yepes H, Monzier M (2007) Seismic, petrologic, and geodetic analyses of the 1999 dome-forming eruption of Guagua Pichincha volcano, Ecuador. J Volcanol Geoth Res 161:333–351

Hall M, Mothes PA (2008) The rhyolitic–andesitic eruptive history of Cotopaxi volcano. Ecuador Bull Volcanol 70:675–702

Hall ML, Samaniego P, Le Pennec JL, Johnson JB (2008) Ecuadorian Andes volcanism: a review of Late Pliocene to present activity. J Volc Geothermal Res 176(1):1–6

Haynes K, Barclay J, Pidgeon N (2008) The issue of trust and its influence on risk communication during a volcanic crisis. Bull Volc 70(5):605–621

Hiltz SR, Diaz P, Mark G (2011) Introduction: social media and collaborative systems for crisis management. ACM Trans Comput-Hum Interact (TOCHI) 18(4):18

IGEPN (2015a) Daily reports of Cotopaxi volcano in 2015 of the Instituto Geofísico de la Escuela Politécnica Nacional. http://www.igepn.edu.ec/informes-volcanicos/cotopaxi/coto-diarios/coto-d-2015

IGEPN (2015b) Special reports of Cotopaxi volcano in 2015 of the Instituto Geofísico de la Escuela Politécnica Nacional. http://www.igepn.edu.ec/informes-volcanicos/cotopaxi/coto-especiales/coto-e-2015

IGEPN (2015c) Informe Especial No. 8 Volcán Cotopaxi. http://www.igepn.edu.ec/informes-volcanicos/cotopaxi/coto-especiales/coto-e-2015/13326-informe-especial-cotopaxi-15-08-2015-1/file

Kumagai H, Nakano M, Maeda T, Yepes H, Palacios P, Ruiz M, Arrais S, Vaca M, Molina I, Yamashima T (2010) Broadband seismic monitoring of active volcanoes using deterministic and stochastic approaches. J Geophys Res: Solid Earth (1978–2012) 115(B8)

La Condamine Ch. Marie (1751) Diario del Viaje al Ecuador, republ. In: 1986, Quito, Publitécnica, 221 pp

La Hora 04/08/2015 (2015) Descoordinación en análisis del volcán Cotopaxi. http://www.lahora.com.ec/index.php/noticias/show/1101849002/-1/home/goRegional/Cotopaxi

Marti J (2015) Scientific communication of uncertainty during volcanic emergencies. In: Loughlin SC, Sparks RSJ, Brown SK, Jenkins SF, Vye-Brown C (eds) Global volcanic hazards and risk, Cambridge University Press, 371 pp

Municipality of Rumiñahui (2015) Contingency plan of an eventual eruption of the volcano Cotopaxi. Sangolquí, Ecuador, p 44

Naranjo JL, Sigurdsson H, Carey SN, Fritz W (1986) Eruption of the Nevado del Ruiz volcano, Colombia, on 13 November 1985: tephra fall and lahars. Science 233(4767):961–963

Newhall C, Aramaki S, Barberi F, Blong R, Calvache M, Cheminee JL, Punongbayan R, Siebe C, Simkin T, Sparks RSJ, Tjetjep W (1999) Professional conduct of scientists during volcanic crises. Bull Volc 60:323–334

Padilla O (2017) Horizontal and vertical evacuation models based on an eventual eruption of Cotopaxi volcano. Unpublished Ph.D. thesis, Universidad de Alcala, Spain

Pierson TC, Wood NJ, Driedger CL (2014) Reducing risk from lahar hazards: concepts, case studies, and roles for scientists. J Appl Volcanol 3(16):1–25

Pistolesi M (2008) Historical analysis and stratigraphy of the post-XII century pyroclastic activity at Cotopaxi volcano, Ecuador. Implication for lahar hazard assessment. Plinius 34:131–138

Pistolesi M, Cioni R, Rosi M, Aguilera E (2014) Lahar hazard assessment in the southern drainage system of Cotopaxi volcano, Ecuador: results from multiscale lahar simulations. Geomorphology 207:51–63
Reischmann T, Toulkeridis T, Aguillera E (2003) Re-awakening of a Volcano: The 3. November 2002 eruption of El Reventador, NE Ecuador. In: EGS-AGU-EUG joint assembly, Abstracts from the meeting held in Nice, France, 6–11 April 2003, Vol 1, abstract #5341
Rodriguez F, Toulkeridis T, Padilla O, Mato F (2015) Economic risk assessment of Cotopaxi volcano Ecuador in case of a future lahar emplacement. Submitted to Nat Hazards Rev
Sigurdsson H, Carey S (1986) Volcanic disasters in Latin America and the 13th November 1985 eruption of Nevado del Ruiz volcano in Colombia. Disasters 10(3):205–216
Smithsonian Institution (1999–2015) Volcano data, weekly and monthly reports. Global Volcanism Program. http://www.nmnhwww.si.edu/
Snailman R (1978) Sangay survived: the story of the Ecuador volcano disaster 1st UK ed. Hutchinson, London: The Random House Group Limited, p 192
Sodiro L (1877) Relación Sobre la Erupción del Cotopaxi Acaecida el Día 26 de Junio, de 1877, Quito, Imprenta Nacional, Ecuador, 40 pp
Stein S, Geller RJ (2012) Communicating uncertainties in natural hazard forecasts. EOS Trans Am Geophys Union 93:361–362
Tanguy JC, Ribiére Ch, Scarth A, Tjetjep WS (1998) Victims from volcanic eruptions: a revised database. Bull Volcanol 60:137–144
Tobin GA, Whiteford LM (2002) Community resilience and volcano hazard: the eruption of Tungurahua and evacuation of the faldas in Ecuador. Disasters 26:28–48
Toulkeridis T, Zach I (2017) Wind directions of volcanic ash-charged clouds in Ecuador – Implications for the public and flight safety. Geomatics, Natural Hazards and Risks 8(2):242–256
Toulkeridis T (2005) Cotopaxi—Una belleza fatal. Documentary, 56 minutes. https://www.youtube.com/watch?v=8Q86QpUuNxE&list=PL4c1zlw7c_2uLL1phcTpH66rEehjxA9n7
Toulkeridis T (2006) New, efficient educative prevention for Ecuador's volcano Cotopaxi. In: Cities on volcanoes 4 (COV4), Quito, Ecuador: 142
Toulkeridis T (2010) Volcanic hazard preparedness in Ecuador. In: Cities on volcanoes 6 (COV6), Tenerife, Spain, p 228 (May–June 2010)
Toulkeridis T (2013) Volcanes Activos Ecuador. Santa Rita, Quito, Ecuador, 152 pp
Toulkeridis T, Buchwaldt R, Addison A (2007) When volcanoes threaten, scientists warn. Geotimes 52:36–39
United Nations (2015) Sendai framework for disaster risk reduction 2015–2030. United Nations, 37 pp
Visschers VH, Meertens RM, Passchier WW, De Vries NN (2009) Probability information in risk communication: a review of the research literature. Risk Analysis 29:267–287
Whymper E (1892) Travels amongst the Great Andes of the Equator. In: New edition with introduction and photographs by McIntyre L, Smith GM (1987) Peregrine Smith Books, Salt Lake City
Wolf T (1878) Memoria Sobre el Cotopaxi y su Ultima Erupción Acaecida el 26 de Junio de 1877, Guayaquil, Imprenta de El Comercio, Ecuador, 48 pp

Part VIII
Paraguay

Chapter 12
An Approach to Geoethics in Paraguay

Jaime Leonardo Báez Presser, Gilda Agostino and Sergio Daniel Ríos

Abstract Developing paleontology, gold mining in the town of Paso Yobai and diamonds exploration in progress, all in Paraguay, are commented on in this presentation. At the approach, each of them was focused by three technicians of different disciplines. This produces ethical sense in common but with diverse feelings.

Keywords Geoethics in Paraguay about paleontology · Gold mining and natural diamonds

12.1 Paleontology

Paraguay, with 406,752 km^2, is a landlocked country located near the center of South America that is divided by the River Paraguay in two natural regions, the Occidental and the Oriental. The country has relatively few geological outcrops, as most of its surface is covered with recent sediments, vegetation (woodlands almost disappeared from the Oriental region of the country in the last thirty years), and anthropic/agricultural activity.

The Paraguayan paleontology, as well as most of the natural sciences in the country, is still greatly underdeveloped, especially in comparison with two of its South American neighbors, Argentina and Brazil, both countries with a long tradition in paleontology. Although the oldest references of fossil discoveries in the actual political borders of Paraguay are dated back to the early nineteenth century (a "Megatherium" discovered in Asunción is mentioned by German naturalist

J. L. B. Presser (✉)
IAGETH NC Paraguay, Gonzalo de Medoza, 334, Gran Asunción, Paraguay
e-mail: jaimeleonardobp@gmail.com

G. Agostino
Minera EXAR S.A., Luque, Paraguay

S. D. Ríos
Departamento de Arqueología y Paleontología, Secretaría Nacional de Cultura, Asunción, Paraguay

R. D. Acevedo and J. M. Frías (eds.), *Geoethics In Latin America*,
The Latin American Studies Book Series,
https://doi.org/10.1007/978-3-319-75373-7_12

Johann Rudolph Rengger in 1835), more detailed references began to be published just in the first decades of the twentieth century, starting with the works by German geologists K. Carnier and J. Schuster, both from 1911 (Carnier 1911; Schuster 1911) and who cited and described, respectively, fossils woods from the "Tertiary" (later discovered to be of Permian age) of Paraguay.

Presser et al. (2004) presented a general view of the Paraguayan fossil record. The fossils discovered in Paraguay to this date correspond mainly to three geological times: the Ordovician–Silurian (485–416 millions of years ago, e.g., Wolfart 1961; Harrington 1972; De Graff 1987), the Permian (298–252 Ma., e.g., Beder 1923; Herbst 1975; Crisafulli and Herbst 2009), and the Quaternary (2–0.010 Ma., e.g., de Bertoni 1924; Hoffstetter 1978; Carlini and Tonni 2000). There are some recent records of Ediacarian (Warren et al. 2011, 2012), supposed Mesozoic (Putzer 1962; Leonardi 1992; De Valais et al. 2012) and Late Tertiary (Morton and Sequeira 1991) fossils as well.

Fossils are considered cultural heritage by Paraguayan legislation (Law 946/82 —"Of protection of the Cultural Beings" and Article 81 of the National Constitution "Of Cultural Patrimony"), considered property of the state and technically protected by those laws. The institution in charge of administrating the Paraguayan fossils is the National Secretary of Culture. In the reality, the situation is very different as still little is done in order to effectively protect fossils and fossil beds or to promote local paleontological research.

The Llandoverian (Silurian) aged San Fernando quarry (Vargas Peña Formation), located near the city of Itauguá in the Departamento Central, is one of the most important examples of a fossil site lacking protection (Fig. 12.1). This clay quarry has provided records of more than forty invertebrate species or morphospecies, many of them first described on the basis of fossils from that place, and published in around twenty articles. It is probably the most well-known Paraguayan fossil site, but the quarry, in private hands, is used (although in low levels in recent years) to made tiles or bricks. There are many quarries, mainly in the Oriental region of the country that are vulnerable to destruction or excessive use. It is, perhaps ironic, that many of the fossils known from Paraguay would not be known without the existence of those quarries, because as mentioned earlier there are few natural outcrops.

On the scientific aspect, fossil collections in institutional repositories are scarce with just two of them being functional, one located in the Paraguayan National Museum (Museo Nacional de Historia Natural del Paraguay and the other in the Geology Department of the Natural and Exact Sciences Faculty (Facultad de Ciencias Exactas y Naturales—Universidad Nacional de Asunción). To this date, Paraguay still lacks a museum or collection that meets all the scientific, display, and educational requirements to be considered truly a proper museum.

On terms of ethical or unethical practices, the existence of hundreds of fossils in scientific collections outside the country, contradicting the Paraguayan legislation, is undeniable. But perhaps, in a more positive side, it must be taken into account that many of them are already published in the literature and "secure" in museums or collections, so they are available for additional studies and waiting for the

Fig. 12.1 San Fernando Quarry in Itauguá, Central Department. *Credit* S. D. Ríos

moment that the country requests their return to Paraguay. This request, however, has to be done responsibly, after the security of their ensuring conservation is attained and not following false and unpractical "patriotic" or xenophobic feelings. However, some effort has been put in order to stop the illegal borrowing of fossils outside the country for either scientific or commercial uses; in these days, there are active collaborations between national and foreign institutions.

Most of the works on Paraguayan fossils published were written by foreign, mainly Argentinian researchers. On the other hand, there is an increase in paper publications or at least abstracts presented in regional or international congresses about materials housed in Paraguayan institutions that hopefully will be long termed.

12.2 Paso Yobai Gold, a Short Summary

The occurrence of gold in Paso Yobai (Guairá, Paraguay) according to versions of the locals was in the middle of the nineties. This gave rise to the presence of informal miners, like countrymen and foreigners. Many farmers, and of other areas of activity gradually abandoned their work to dedicate silently to this new line, hoping to get rich from the overnight.

Artisanal mining soon became an important source of revenues for miners, especially after the emergence of the Canadian mining company (Latin American Minerals or LAMPA S.A. in Paraguay, between 2003 and 2007), whose installation in the place made that moved important flow of money into local labor. In parallel the fact that a multinational company has been installed in place with a large investment attracted the attention of many foreigners who were adding to the already growing informal mining so that, in practice, until today government actions did not exist because their jobs do not have a legal form, they do not pay royalties, taxes, they do not do social work, etc., and most end up working in difficult conditions, in constant danger, just saving the fresh bread daily and rubbing shoulders with the use of mercury, whose health consequences are long term, making them seem unconscious of the real danger to both health and environment.

Since the Canadian mining company was installed in Paso Yobai, it established a program of relations with neighboring communities to the project. In this context, it has been working covering several important issues such as health, education, the environment and among others, communication activities that the company has been doing.

Their presence has made possible an important economic movement locally. With very positive political, socially it has been positioning itself in the community to the point that it has become part of it. However, in recent years, it has been facing many difficulties on several fronts and everything that was achieved has decreased to the point of losing the position that once gained with little popular events, including the dismissal of workers and suppressed their community relations area, which in the absence, the company lost its human face.

Meanwhile, small-scale miners have also been eroding its work, since despite its rich empirical experience, limited technical expertise has led to stagnate to reach excavation depths in front of mine who can no longer manage. Gold has been fading from their hands remaining in the "pits" as an indefinite lethargy (Fig. 12.2).

In Paso Yobai, mining caused many families to change their lifestyle in great shape. Many have taken advantage of the economic prosperity to give greater impetus to its objectives in a very positive way, while others have lost everything or almost everything, thinking that gold always sprouts with the same momentum as when it appeared. Everything is good, when common sense and goodwill are used as tools for human ethic development.

12.3 Diamonds

As indicated on the website of the American Museum of Natural History about the diamonds: "Born from billions of years of crushing geological force, the diamond is a truly extraordinary material". And in Berkeley University classes (http://nature.berkeley.edu/classes/eps2/wisc/diamondcom.html, accessed in August 2015) is commented that: "Diamond, one of the world's most important mineral resources, is made of pure, natural carbon with the atoms organized in a close-packed cubic

Fig. 12.2 Arroyo Itá stream sediments rich in gold (1–2 mm) collected by Jaime Presser in 1997. *Credit* J. B. Presser

arrangement that gives the stones their hardness. The external forms of natural diamond crystals (isometric system) show the same symmetry. The common crystal form is the octahedron, which looks like two four-sided pyramids placed base to base. Because diamond is so much harder than any other natural or artificial substance known, it is ideal for both gem and industrial purposes. Special optical properties guarantee its preeminence among gems. First, its high refractive index (2. 418–2.417), or light-bending ability, enables it to throw back almost all the light that enters a well cut gem. This gives rise to the gem's brilliant, or adamantine, luster. Second, it exhibits strong dispersion (0.058) or the ability to separate the various colors of the spectrum. This causes the gem to throw back the bright flashes of separated colors ("fire") for which it is particularly noted."

Diamonds occur in two general types of deposits worldwide: volcanic pipes, also known as kimberlite (kimberlite, orangeite and lamproites and occasionally also in laprophyres and other alkaline rocks) pipes and alluvial, or placer, deposits, which were formed by the erosion of diamond pipes over millions of years. Diamonds have been mined for many centuries in India and, later, in Brazil which were the two main sources of the precious stones until the South African discoveries. South African discoveries with one initial phase of diamond rush which saw 10,000 prospectors converge on the Vaal River between 1869 and 1870. And, later it is given the biggest step in diamond mining—the discovery of the diamond-bearing kimberlites: Du Toit's Pan, Bultfontein, and Kimberly (1870/1871) (Tappert and Tappert 2011 and this work).

Only a small proportion (~10%) of the more than 5000 known kimberlites is diamond bearing. Among those, only a small percentage (<1%) contains diamonds in sufficient quantities to permit large-scale industrial mining. These few diamond-rich kimberlites, however, supply the bulk of the world's natural diamonds, which on an annual basis equals around 30 metric tons, which is equivalent to around 150 million carats (Tappert and Tappert 2011).

As well as according to a paper britng by Pisani (http://www.cnbc.com/id/48782968, accessed in August 2015), "the worldwide retail market for diamond jewelry was $60 billion in 2010. Because large, commercially viable diamond mines are a rarity, today, there are only about 20 major diamond mines in the world. And the big supply is even rarer—only 11 mines make up 62% of the world's production of diamonds by carat! And they are getting even less common. The last major diamond mine was discovered in Zimbabwe in 1997. No matter how big the mine is, you have got to move a lot of rock to get at a very small number of diamonds. Production varies by mine, but the world's "richest" mine—the Jwaneng mine in Botswana—has to move a ton of rock to get 1.4 carats of rough diamond, on average. That mine moves 8 million tons of rock a year and sells the rough diamonds for an average of $134 a carat. That is $1.5 billion in revenues, from a single mine. And it produces profit margins of 24%."

12.3.1 South America Diamonds

The vast majority of natural diamonds are formed in the Earth's mantle at depths of more than 140 km. Within the mantle, diamonds form primarily in the rigid part of the upper mantle, which is referred to as the lithospheric mantle. In addition to high pressures, diamond formation is favored by relatively low temperatures. In nature, these conditions restrict diamond formation to regions with a very low geothermal gradient. Regions with such low geothermal gradients are generally confined to the geologically old parts of continents, which are referred to as cratons. Cratons are composed of crustal basement rocks, such as granites and gneisses that are more than ~1.5 billion years old. The lithospheric mantle roots beneath cratons can extend to depths of more than 250 km, and due to their relatively low temperatures and extreme thicknesses, they are the main locations of diamond formation in the Earth's mantle. Diamonds are believed to crystallize primarily from carbon-rich fluids that percolate through these lithospheric roots (Tappert and Tappert 2011). Throughout Earth's history, kimberlites/lamproites/lamprophyres originating from deep within the mantle passed through the lithospheric craton roots and during their ascent, the magmas incorporated the diamonds.

In Brazil, diamonds were discovered in 1714. From 1730 to1870, Brazil was the world's leading source of diamonds. Between 1890 and 1901, secondary diamond deposits were discovered in eastern Guyana and Venezuela. Records are also known in Colombia, Uruguay, and Paraguay. Alongside Brazil, Venezuela is today one of the largest exporters of rough diamond of South America. In Brazil the kimberlites search started at the end of the 1960s, from 1969 to 2010 Brazil would have been recorded over 1500 kimberlites and some with diamonds but all without representative volumes for economic exploitation, except for a small body of low grade diamonds in the state of Bahia, the Brauna Kimberlite, which will be operated and is estimated to have a useful life of seven years. Kimberlites also were also discovered in Venezuela, Paraguay (lamproites and kimberlites with diamonds, Presser et al. 2013) and Uruguay. Paraguay, Argentina, Uruguay, and Brazil share the area with thick craton (200–300 km deep = cratonic environment ideal, Presser unpublished data)—The Rio de la Plata craton, believed to be the target of future discoveries of diamond-bearing kimberlites that could be framed as of economic importance.

12.3.2 "Blood Diamonds"?

Diamond is also shown to be a highly lucrative business, the late twentieth century was subject of prejudice by the international community, and this occurred when some African diamonds (from Angola, Liberia, Sierra Leone, etc.). They were called as "blood diamonds" or "conflict diamonds." This time of "conflict diamonds" gave birth to the Kimberley Process (KP). The KP is a joint governments, industry, and civil society initiative to stem the flow of conflict diamonds—rough diamonds used by rebel movements to finance wars against legitimate governments. The Kimberley Process started when Southern African diamond-producing states met in Kimberley, South Africa, in May 2000, to discuss ways to stop the trade in 'conflict diamonds' and ensure that diamond purchases were not financing violence by rebel movements and their allies seeking to undermine legitimate governments (http://www.kimberleyprocess.com/ accessed in August 2015).

Fifteen years passed (2015), and ingredients that support the conflicts between governments and between countries in Africa, Asia, and the Middle East found the "fossil fuels" as a means of financing. Perhaps one imagined into carry out something similar to KP for oil? Or there is no conflict oil? The society is very special, its ethical values not in few occasions they are of "double moral". We have the impression that there are more people in the world rejecting diamonds by prejudging that they would be "blood diamonds" that people wondering if the fuel in their vehicles or cooking gas in their homes would be of "conflict oil."

On the other hand, where government presence is real and there are kimberlite diamond mines, the present of "conflict diamonds," in practical terms, is nonexistent. Brazil and Venezuela are more marked by the extraction of alluvial diamonds. Paraguay and potentially Argentina and Uruguay as the mining spirit of the diamond gets maturing will face also real alluvial diamond extraction (Fig. 12.3). This is the type of mining that is more thumbed by illegality; or for saying in a phrase thumbed through the "conflict diamonds"; but could be minimized or avoided with effective and ethical government policies.

Fig. 12.3 Picture of the two largest diamonds already found in Concepción (Paraguay) 3.33 and 2.65 carats. *Credit* J. B. Presser

References

Beder R (1923) Sobre un Hallazgo de Fósiles Pérmicos en Villarrica (República del Paraguay). Boletín de la Academia Nacional de Ciencias en Córdoba 27:9–12

Carlini AA, Tonni EP (2000) Mamíferos Fósiles del Paraguay. Cooperación Técnica Paraguayo-Alemana. Proyecto Sistema Ambiental del Chaco-Proyecto Sistema Ambiental Región Oriental. La Plata, Argentina, 108 pp

Carnier K (1911) Über das Alter der Hölzer aus dem Randgebiet von Villa Rica in Paraguay. Mitt Geogr Ges 6:430–431

Crisafulli A, Herbst R (2009) Gymnospermous wood (Coniferales, Taxales and Ginkgoales) from the upper permian tacuary formation, Eastern Paraguay. Paleobiodiversity Palaeoenvironment 89:95–109

de Bertoni AW (1924) Sobre mamíferos fósiles del Paraguay. Revista de la Sociedad Científica del Paraguay 1(6):93

De Graff AE (1987) Estado actual del Silúrico-Devónico en el Paraguay. Informes Científicos—Instituto de Ciencias Básicas 5(1):34–45

De Valais S, Filippi V, Molinas S, Souberlich R (2012) Descripción de una huella de terópodo de la Formación Misiones: primer fósil mesozoico fidedigno de Paraguay. Andean Geol 39(3): 541–547

Harrington HJ (1972) Silurian of Paraguay. In: Berrey WB, Boucot AJ (eds) Correlation of South American silurian rocks, vol 133. Geological Society of America, Special Paper, pp 41–50

Herbst R (1975) Helechos Pérmicos del Paraguay. In: Actas I Congreso Argentino de Paleontología y Bioestratigrafía, vol 1. Tucumán, Argentina, pp 420–438

Hoffstetter R (1978) Une faune de Mammiferes pléistocénes au Paraguay. Comptes Rendus Sommaries des Séances du Societé Geólogique du France 1:32–33

Leonardi G (1992) Sulle prime piste fossili di tetrapodi del Paraguay. Paleocronache, pp 66–67
Morton LS, Sequeira PA (1991) Pelecípodos de agua dulce de la Formación Ituzaingó (Plioceno Tardío) de la presa de Yacyretá, Departamento Itapúa, Paraguay. Revista de la Asociación de Ciencias Naturales del Litoral 22(1):25–34
Presser JLB, Buongermini E, Fillippi Amábile VO, Fernández Crossa V, Báez Almada AB, Zarza Lima PR, Oporto Migone O (2004) Algunos antecedentes paleontológicos del Paraguay. Boletín del Museo Nacional de Historia Natural del Paraguay 15:95–110
Presser JLB, Molinas Gini M, Franco González O, Céspedes Allende JM, Cantero Cantero J (2013) Paraguay: una nación diamantífera. Boletín del Museo Nacional de Historia Natural del Paraguay 17(1):5–19
Putzer H (1962) Die Geologie von Paraguay. Beitrage zur Regionalen Geologie der Erde, Bd. 2. Berlín, 183 p
Rengger JR (1835) Reise nach Paraguay in den Jahren 1818 bis 1826. Aarau
Tappert R, Tappert MC (2011) Diamonds in nature: a guide to rough diamonds. https://doi.org/10.1007/978-3-642-12572-0_1. Springer, Berlin Heidelberg, 142 p
Schuster J (1911) Osmundites von Sierra Villa Rica in Paraguay. Deutsch Botanische Gesell 24 (7):534–539
Warren LW, Fairchild TR, Gaucher C, Boggiani PC, Poiré DG, Anelli LE, Galeano Inchausti JC (2011) Corumbella and in situ Cloudina in association with thrombolites in the Ediacaran Itapucumi Group, Paraguay. Terra Nova 23:382–389
Warren LW, Pacheco MLAF, Fairchild TR, Simões MG, Riccomini C, Boggiani PC, Cáceres AA (2012) The dawn of animal skeletogenesis: ultrastructural analysis of the Ediacaran metazoan Corumbella werneri. Geology 40:691–694
Wolfart R (1961) Stratigraphic und Fauna des alteren Paleozoikums (Silur-Devon) in Paraguay. Geol Jahrb 78:29–102

Part IX
Peru

Chapter 13
Geoethics: The Peruvian Case

H. Saul Pérez-Montaño

Abstract The term geoethics is a new word in Peru; however, it has been using in practical terms by the old Peruvians (Incas), because they were promoting geoscience taking care of the environment and society centuries ago. This beautiful concept has changed along the time, although there are places especially in the highland cities where people continue following this ancient tradition. Besides the economic, political stability, and the recent support of research and science, there is a separation between the concept of geoethics and the perception and work done for many institutions especially when we talk about the preservation of land, water sources, and environment. Although there are important advances and some policies have been established to promote the equilibrium of the economic activities and the conservation and preservation of the natural resources and land. In the current context, there is a lack of qualified professionals with enough background on ethics, these new professionals should take the lead and be responsible to promote a new educational cultural in geoethics with a vision of social responsibility.

Keywords Peruvian case · Geoethics · Geology · Education · Ethics Social responsibility

13.1 Introduction

The meaning of geoethics is not a recent topic in Perú. The old native Peruvians, founders of the Tawantinsuyo Empire (Incas Empire), were very familiar with ethic rules, then people could remember in their salutations every day; *ama sua* (do not steal), *ama llulla* (do not lie), *ama quea* (do not be lazy). These simple rules were

H. S. Pérez-Montaño (✉)
Universidad Católica San Pablo, Urb. Campiña Paisajista
S/N Quinta Vivanco, Barrio de San Lázaro, Arequipa, Peru
e-mail: hsperez@ucsp.edu.pe

R. D. Acevedo and J. M. Frías (eds.), *Geoethics In Latin America*,
The Latin American Studies Book Series,
https://doi.org/10.1007/978-3-319-75373-7_13

closely attached to their activities and normal lives and were incorporated to their activities related to the land, environment, and geosciences. Incas promoted collective work (Minka and Ayni); so once a year for a specific time people went to help on different needs, such as work on the land, mineral extraction, metallurgical artisans, climate disasters, or construction of the gorgeous and magnificent Andenes, an example of how to prevent erosion on the soils especially in the Andes and other needs of the empire. This cultural concept has been passed through the years in some of the highlands cities. These activities were designed to take care of the environment and Earth, and this meaning was deeply attached to the Peruvian culture because of Incas considerate Earth as their mother (Pachamama). Although this extraordinary conception of human beings and our planet has been forgotten partially, there are still some places on the highest lands of Peru where this tradition has been kept (www.wikipedia.org/wiki/Inca_Empire).

13.2 Current Situation

Since the last decade, Peru has been enjoying political stability and sustained economic growth; according to the Ministry of Foreign Affairs Peruvian Minister for the period 2015/2016, the significant percentage came mainly from mining operations having a positive repercussion on the Peruvian economy. This positive impact was related to mining extraction; thus, Peru is the third large producer of copper and zinc, as well as in the ten list producer of gold, silver, and other minerals. Regarding land, Peru is the third largest country in South America where the coast is a desert with an area of 10.7% the highlands contain the 31.8% here is where the mining operations have placed and the jungle where are localized the significant deposits of petroleum and natural plants with 57.5% (BCRP and INEI, www.ey.com).

An overview of the main activities shows that the mining operation contributes largely to the economy of the country, and they have been promoted largely by the Peruvian Government. However, there have been considerable concerns about the impact on the environment of the mining and other productive activities. Although mechanisms of control have established, there is a gap at the time to apply all of the legal requirements regarding protection of the environment. This convoluted situation with a lot of issues has become even more complicated because of the limited institutions that work with scientific bases and the small number of professional and scientist converging in organizations that can contribute to geoethics.

13.3 Institutions Related to Geoscience in Perú

The first institution that has activities in Peru related to geoscience was INGEMMET; it had their origins in 1852 when the Central Board of Engineers was established having a base foreign and national engineers hired from Europe. The establishment of this organization had coincidence with the presence of Antonio Raimondi, who was hired to make the first map of Peru, showing natural resources, geographic distribution conveying an enormous amount of work and being for decades the main reference in the geoscience field on Perú (www.ingemment.com). In the line of time INGEMMENT has been having different names, however, since 1979 changed its name to the current name. In 2007, was made the last change adding the responsibility of administrative jurisdictional and geoscience because of the previous duplicity with other governmental organization (www.ingemment.com).

The other institution that plays a pivotal role in the development of the geoscience in Perú is the Geological Society of Perú (GSP), this organization was founded in 1924 and two years later was officially recognized as an official society by the Peruvian Government, after years of diligent work to establish geoscience as important field of research this institution promoted and participated actively in the formation of Peruvian Institute of Geology later named INGEMMET. This society provides guideless to make the prospection and exploration of naturals resources and hydrocarbons. An important contribution of the SGP are the studies on deglaciation in the Andes Mountains and their consequences, the vulnerability of the glacial lakes at the White Mountains of the Andes, both topics are related to climate change (www.sgp.org.pe).

A remarkable fact was that GSP was making call since 1949 to publish thesis on Geology, and since 1954 has been participating as an organizer of Peruvian's congress related to geology field. These activities are critical because they provide the opportunity for potential research from pre-grade, and professionals to show they interest and work done in the geoscience field (www.sgp.org.pe/nosotros/historia/).

The institutions that form professionals in geology currently there are ten universities in total that offer the program, two of them are in the capital on Perú (Lima) and the rest of them in different departments as Cerro de Pasco, Cajamarca, Piura, Arequipa, Cusco, Puno, Tacna, and Ancash. Although there are not directly related to geoethics, these institutions have the responsibility to form the professional that are involved directly with geoscience that involve lithosphere, hydrosphere, atmosphere, and biosphere, in our case to be more specific the formation of the Andes, the huge fluvial natural resources available in Perú, and the richness of our Jungle and their diversity.

13.4 Peruvian Geography and Education Efforts

The richness of the Peruvian Geography is huge and there is an agreement on it; however, the gorgeous geological formations can be a source of a lot of troubles because of the climate and other weather conditions that demands organization, education on different levels, high level of compromise from our youngest professionals and commitment of the local, regional, and national authorities on high positions and the power of decisions, in order to be aware, and prepare adequately to the most vulnerable population and cities in our country. At this point, it is important to recognize that we are starting and making progress small progress since 2011. However, the number of the works from basic research from the universities has shown enough progress, and the conclusion is that a lot of work is needed. Besides the latest economic support from the government to science through CONCYTEC, INNOVATE PERÚ, and the Minister of Production there is a lot of work to be done regarding geoethics.

One of the major efforts make for the government and the education authorities was the promotion of the basic and applied research; as a country it is important to promote that our students could understand the limitations and potentialities that we have about science, learning the basics about the scientific method, critical thinking, formulating solutions about complex situations among them the understanding of geoethics. In order to promote and fill the gap that we have in the geology field, since 2011 Perú has been making huge efforts to promote science and research in geology field in Fig. 13.1, the data show data from 2002 to 2017, and this data indicate an increasing number in terms of research made by the universities. An important change in this numbers is that since 2013 master degrees have been publishing and this tendency has been observed along the time, and also we can observe a continuous work in the geology field (www.renati.sunedu.gob.pe).

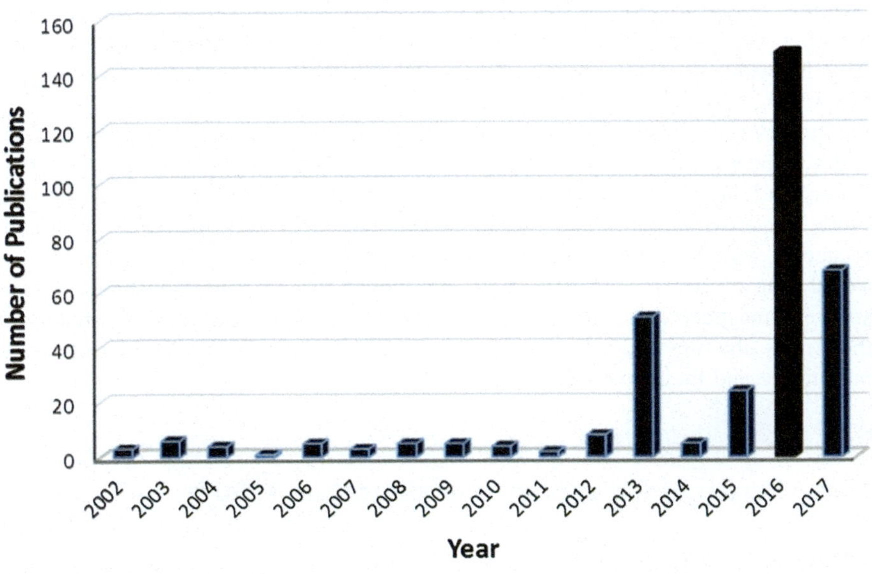

Fig. 13.1 Growth of scientific publications in Perú Modified from http://renati.sunedu.gob.pe

13.5 Conclusions

Perú has a long tradition as a country with friendly relation with the environment. However, the current situation of geoethics in Perú needs to be considerate carefully. A positive influence in the economic situation of the country there has been in the last decades as a consequence of the mining operations. Other important factor that contributes positively to this situation was the political stability. However, there are important aspects in geoethics that need a lot of work. Another sector that has been benefited of all these positive aspects is the educational sector, there have been recently a support to promote basic and applied research from the government, although these efforts are on the initial level, progress has been noticed and all the institution involve on the field directly and indirectly have to make an effort an collaborate to create the bases of an unified document and the bases of an ethics policy for Perú. An important aspect that explains this situation are the few institutions working with geoscientist that understand the importance of the social aspects, this part is critical because geoscientist with ethical dimensions will play a fundamental role in the society and will develop an educational culture and geoethics norms.

Part X
Venezuela

Chapter 14
The Epistemological Evolution… The Beginning of a "Geoethics" Cosmovision

Gladys Dávila and Robert Bello

Abstract The objective of this article is to provide the reader a view of the link between the Geoethics on the geological knowledge management and the energy scope of Venezuela and Latin America from the epistemic point of view, some theoretical elements of the two mentioned events and recent contributions about those elements in convergence. From the literature review, this work deals mainly with Potter (Perspectives in biology and medicine, 1971), Habermas (2009), Morin (Más allá de la globalización y el desarrollo Gazeta de Antropología, 2003), Nikitina (Geoethics: theory, principles, problems, 2016), Martínez (2008), and Bautista (Jornada de divulgación del programa de investigación y I encuentro de investigadores, 2016), Bello (Gestión de la innovación en el contexto organizacional inteligente: visión epistémica desde el ámbito bioético/transcomplejo, 2013) and Rivas (Liderazgo bioético: una cosmovisión sistémica en las organizaciones del sector energético, 2017), Padrón (Revista educación y ciencias humanas, 1998, Una visión evolucionista y cognitiva de la producción de investigación, 2016), on a nowadays view of the "Geoethics," the Bioethics valuation, the reflexive actions of the human being, and the emergent paradigms for the knowledge production, and driving actions that cause the fundamental changes that the humanity is demanding. The assumed methodology, as it is stated by Bautista (Jornada de divulgación del programa de investigación y I encuentro de investigadores, 2016), conducts to steps of complex construction in an also complex path, allowing holistic "persovisión" of the elements that converge in both the energy sector and the academic, where human interactions in society are positively oriented toward the use of the planet resources inviting to the construction of an own epistemic construct from the reality of Latin America that arrives to a sustainability that assures the life and human survival. The transcendence of this procedure considers not only the necessity of deep transformations at the internal levels of the human being but also at the community environment, as well as at

G. Dávila (✉)
PDVSA, Maracaibo, Venezuela
e-mail: riolitas@hotmail.com

R. Bello
CANTV, Maracaibo, Venezuela

R. D. Acevedo and J. M. Frías (eds.), *Geoethics In Latin America*,
The Latin American Studies Book Series,
https://doi.org/10.1007/978-3-319-75373-7_14

the organizational structures of large-scale, for the recognition of the man as a being that is universal and part of the global system.

Keyword Epistemology of geoethics in latin america and venezuela

14.1 The Evolution of the Ethics, Bioethics, and Geoethics within the Energy Sector

The intellectual processes of scientific production/technology stay at the complex thought a systematization and epistemological relationship of the current status of the researching. This essay shares a view that contains the extension or complement of elements that get integrated in a reality to be studied, contextualized in an interpretative and reflexive way, in the thought of the complex paradigm.

The explorer approach supported on the perspective of the researcher stays a contribution when put together the synergistic comprehension of the nature at all dimensions, biotic/abiotic, with the perspective of the observer that merges the significance of the tangible and intangible through the interaction of the complex phenomenon. In other words, the systematic conditions represented within the scope of thinking translate to a perception of the knowledge with a reflexive relevance that leads to an integration of the human being with all the daily manifestations.

The context of this article converges between the realities perceive at the academic organizations and the energy sector in Venezuela, considering phenomena like the huge volumes of hydrocarbon reserves of the nation, the recent findings of gas reservoirs, the potential of the mining arch located on the southeast area of Venezuela, the fresh water reserves, the threatened reality of the Amazonian forest of the country, the natural risks, as well as the great human resource potential technically prepared in national universities in areas such as engineering and related careers, where the geologist is involved in.

Those elements demand synergistically a Geoethics formation in both areas: Academy and organizations manage the knowledge of human talent, not only to preserve the national sovereignty but also to power a hope of human survival as a response to the exhausted traditional epistemic systems that only allows a reductionist view within the positivism.

In accordance with the above statements, the reality is a social construction that proceeds from the person that perceives the world with the biological configuration of its mind with its cognitive base, in other words, with specific approaches or point of view… There are some authors that propose the duality of the reality (both sides of the coin, the mirrored reflection) or complementarity, and other authors that propose the reality like a structure of three elements (mind/reality physic/ interrelationships with the others), in any case, does not escape from a cosmovision of complex approach.

On the other hand, the perspectives of the persons converge on analyze the world, assimilating the research and the science with an explanation that exhibits a relational structure, paraphrasing Padrón (2016), notorious between the *Odgen* triangle and the three worlds of *Popper*. The *Popperian* thesis of three worlds assumes three “realities”: the world of the objective things (it involves everything that we capture through our senses); the world of the subjective contents (comprises the consciousness contents and the internal life of the subject); and the world of symbolic-cultural constructions that transcend the individual to be placed at the society domains (ideas and collective representations such as idiom, religion, art, science, law, and others).

Thereon, Padrón (2016) states that in addition to the human responses, events or situations obey to a series of knowledge gained by natural learning, generational and induced through the study and the research carried out. In the society and the organizations exists the need to create and give dynamism to the knowledge in order to systematize processes and evolution to the development. The organizations utilize the information systems and the technologies for this purpose, generating networks which are becoming better interconnected, with the objective to create personal/organizational knowledge, connoted in linked aspects from the root to the human mind condition (on the bio-cognitive configuration) and its way of perceiving the reality.

The complex thinking that Morin defends (2003), mentions the human passage through the planetary era. The effects of the globalization, on a convenient understanding to define “the sustainable development”, masks a real devastation of the natural resources that make possible the human life on the Earth. Due to the evident fight against the desertification of the inhospitable territories and other serious problems for instance the depletion of the resources of economic value, it is proposed to relaunch the dream of good living or good life, simply enjoying the simple thinks that make humanely the human life, and thus the life of the other organisms in the ecosystem.

Assimilating the above statements, the Theory of the Systems, according to Morin (1998), stablished the bases of a thinking about the organizations, indicating that the whole is more than the sum of the parts. In the meaning of the significance or replication of new elements, that be born from the organization as a whole and retroact simultaneously over the parts, enhance a relevant conceptualization over the contraposition that points out the whole in minority with regard to the sum of the parts, because the peculiarity possess systematic features inhibited by the organization as a whole.

Reflection constitutes an action of the thought where all the ways converge to perceive the reality as a system of feedback where thinking about the science allows the cumulative of the knowledge, that progresses and renew on its epistemic basis, it consolidates a reference with the behavior of the nature, as well as the management models of the geological knowledge about the local and global geosystems and the harmonic equilibrium of interests of all the social groups that addresses the use of the mineral resources and useful properties of the underground or geologic environment, through the requirements analyzed with a consecrated model from the systematic perspective, observing the linked aspects on an integrated way: nature, humans, organizational and socials.

In this regard, Morin (1998) raises about the "hologrammatic" principle an existing definition in the apparent paradox of certain systems on which not only the part is on the whole, but also this whole is replicated or represented by this part; as an example, he mentions an event associated with the totality of the genetic resource presents on each individual cell. Similarly, the human being contemplates a part of the society, and at the same time, the society is present in each individual reflected through the idiom, culture, and rules.

The human relationships have its expression in the society, as a reflect of each culture at each time, according to the sociopolitical, economic, and moreover environmental context. It is therefore that the empowerment of the science, the technology, the communications, in the current vulnerable and corruptible environment, consider a technologic and economic development in parallel with a weak or null reflexive responsibility and also a null or fragile axiological relevance; on this context, the organizational elements demand deep actions that carry out not only the founding Geoethics components of biggest size and weight but also the light, flexible, and tiny ones be able to move further within the space-time scope in order to assure hologrammatic projections of proactive scenarios.

In the same category of ideas, Frederick (1995) mentions that the Fractal Theory proposed by Benoit Mandelbrot (1975) applies for the study of elements that in mathematic terms does not have an entire spatial dimension, but fragmentary, that is why the social phenomena, linked to the psyche, the culture, the socialization, the communication, and the organizational behaviors, allow a fractal representation like the proper nature elements.

Additionally, based on the fact that the planet Earth is part of the natural world, and its study comprises not only the Earth but also by what it is made up of and surrounding, including the interaction of the human and its elements, makes the fractal event an imbued assimilation in the complexity with an approach that leads to research/understanding.

On the other hand, Geoethics considers an animated nature that is reproduced by selection processes and also by an abiotic nature that is reproduced by fractal effect regulated objects as copies of others, components of a single system in continuous learning evolution.

However, the man on his making-ideas process, he reflects the world order and vice versa, thus, the addition of any individual human action impregnates by coincidence an impact to his environment; therefore, assuming proactive actions that feed ideas and an assertive world for living, allowing a reception of results equally assertive prone to a replicate at the global social level.

In consensus with the above paragraph, Migueles (2011) indicates that according to the theory of the dissipative structures of Prigogine, any open system is dynamic because it involves interaction with the surrounding; in the nature, nothing is fixed, everything is in continuous motion, generating metaphorically a convergent reference where a rock is still a continuous dance of subatomic particles, which intensively dynamite its experiences. This assurance expresses that the study scale of the plane is such a broad and complex, which admits the existence of new integrative disciplines and invades in emerging the assimilated thinking in the absolute objectivity of the academic order.

The understanding about the transdisciplinary that states Nicolescu (1996) involves the connective spaces between disciplines, in other words, the dialogical as an input that allows to execute the transdiscipline; raises that the reality has a trans-subjective dimension; similarly, it is composed by a quantum dimension and a macrophysics dimension. In addition, the consulted source shows that the way on which the human generally perceives the real environment ignores the perspectives of multidimensional and multi-referentiality.

In this respect, the quantum physic and the cybernetic oppose concepts of exclusive peers that apply to many elements, excepting time, to which apply the entropy of the disorder, confirmed with the nature of the universe. These assev- erations assume that the "invariance" seen at microphysics level that makes possible the universe we know and see is composed of matter, and opposite effect must be scientifically proven: the existence of an antimatter.

The transdisciplinary concerns, as the prefix "trans" indicates, what is at a time between disciplines, through the different disciplines and further of the complete discipline. The objective of it is the understanding of the present world on which one of the imperatives is the unity of knowledge (Nicolescu 1996).

It is deducted that what is reduced by the classic thought to facilitate the expertise in each discipline allows to assume that they circulate in isolation; however, those spaces between the disciplines are not empty, as it is conceived by the classic physic; in contrast, the transdisciplinary admits interactive connections and languages between disciplines breaking the limitations of the classic thinking.

The transdisciplinary view proposes us to consider a multi-dimensional reality, structured at various levels, that replaces the one-dimensional reality, at only one level, of the classic thought. Nicolescu (1996) shows the complementary connec- tion of the different levels of the reality conception, without relevance of some over others; the previous statement, contextualizing the natural world at those levels that are fully consistent, in the same way, the flow of information stated in the globality of the everyday horizon, is transmitted between the different levels of the reality of our physic universe decanting into the knowledge evolution which is always open.

The non-empty zones between the disciplines, "zones without resistance to the perception," constitute the transdisciplinary subject. Those zones communicate to the transdisciplinary subject and object, in a way that they found a place to con- jugate both studies: the universe and the human being, supporting each other. In this respect, the author states that the "transdisciplinary model of the Reality has, more particularly, important consequences on the study of the complexity," circumscribes the founded language about the inclusion of the third that is found between on "why" and "how to do it," between the "who"? and the "what"? assuring the user language from the transdisciplinary language. "In the transdisciplinary vision, the complex plurality and the open unit are two facets of only one and same Reality" (Nicolescu 1996).

In this confluence of ideas, epistemically, the transdisciplinary dialog facilitates the approach and solution of new problems and constitutes a fundamental element for the study of complexity, this dialogic allows to comprise problems related to not only what is called hard science but also to the soft ones, and it guarantees the transcendence of the created knowledge.

About this subject, Morin (2011) indicates that at the crossroads of destruction in the man, there exists an alternative of conquering a planetary well-being that conducts to a human well-being, and this alternative starts with a turning to the think how and performs social reforms about the art of good living well through a metamorphic process.

These transformations on the thinking how, researching, producing knowledge, objectifying knowledge imply a holistic and integral modification in every human sense, about the use of food, raw material, medicines, technology, transportation, communications, law, policies, for its transcendence from a planetarium and a local domain.

However, the analysis of the social implications of all the scientific knowledge in evolution circulates in logos narrative/discursive/ethics/aestheticizing, relating to not only a reflexive/critic gesture but also with the complementary analysis of the recursiveness, dissipative structures of the systems, supported by the thermodynamics and no-liar thermodynamic, the chaos, the networks, fractals, variance, and the internal entropy of the universe itself, among others. Connoting this perspective, to do science does not exclude elements, tasks nor actors, and in contrast, it is an interdisciplinary and transdisciplinary construction.

On the other hand, the possibility of being in an open epistemology allows for uncertainties and the dialogic that enables an important progress to the knowledge. For Morin (1998), the dialogic is the fundamental idea on the complexity; it permits to connect the order/disorder/organization. Beyond dichotomy statements or antagonism, "the dialogic principle allows us to keep the duality within the unit. It associates two terms at the same time: complementary and antagonist" (Morin 1998).

In consequence, the aim of these changes occurring from the interventions on the scientific knowledge production is to transcend; therefore, the transdiscipline applies to this invitation to generate the Geoethics construct.

According to the proposed theory, the principles, such as the systemic, quantum, of complementarity, the fractality, the chaos, the computing, the cybernetics, the auto-organization, as Bautista indicates (2016), are common elements of a indissoluble cosmogonic singularity that is what we called complexity, and nowadays, it constitutes a paradigm that decants into open methods to the permanent construction and breaks the previous reductionist structures.

Regarding it, the geology as the science that studies the Earth results in a connective tissue that does not discard the foundations of other sciences that consolidate the analysis of the tangible and intangible, and historically, since the end of the nineteenth century until the earliest XXI century, the Empiric or Inductive Analytic or Positivist trend determines an important part of the context of what is called Natural Sciences.

The research scientific process refers to a broad concept that encompasses all the sciences, from pure sciences until the socials and humanistic, moreover, nowadays, when the researcher is reflexive aware and considers all the intrinsic variables of the human being, not only the quantifiable elements. Therefore, the emergence of new realities through the time causes the arising of new alternatives in the knowledge construction, and in this respect, the following is quoted: "The research is considered a human activity, oriented to collecting new knowledges and its applicability to solve problems or questions of scientific character" (Padrón 1998).

At a broader level, it is disclosed the emergency of new epistemologies; however, it includes the subject and the influence of social/contextual factors in the scientific process of investigation. This is translated in the thinking styles and the criteria to generate the classes of the epistemological approaches.

With the introduction of new points of view, the evolution of the epistemological approaches is analyzed; however, Padron (2007) reveals, in regard the recent trends in the epistemological development, that they obey the emergence of new problems, the restatement of ancient problems, the new proposals of solution, and the new ways of exploration.

The bioethics does not escape of this dilemma, and it is crucial its emplacement at the social scale of the geologic work, whereby, between the emergent paradigms for the knowledge production, erupts the complex "cosmovision" where flows the linkage of the stated events in this reflective proposal to be involved in the generation of a proper epistemology of Latin America realities.

The bioethics is the result of "a movement, an interdisciplinary work, a growing process of searching moral values and as such it is required to give time to take the form, sense, method and regulation. It is not a matter of a regulation already made, coming from outside with coactive character, but about something that is progressing and surely it will obtain the character of a multidisciplinary science, multifunctional and poly-qualitative with the effort of everyone" (Navarro et al. 1999).

In Latin America, in the last few years, researchers are dedicated to examining the bioethics from different point of views: attitude, social responsibility and social rights, laboratory investigations about regenerative medicines with human being, and also in subjectivities, law for life protection, natural disasters, among other challenges of the human activities in his own well-being.

The perspectives of the bioethics for Potter are defined by three views, the global, bridge, and deep, corresponding to an evolution of the common goals. In this sense, the bioethics bridge consists of the link between the biologic science and the ethics; in this way, it refers to the survival of broad scope of the human being, in a decent and sustainable civilization, requiring the development and maintenance of a social/ethic system.

In that context, the bioethics is conducted through a bridge of future knowledge, in a sense of coexistence system with the nature, in addition to the recognition of the constant human errors, in order to pursue the wisdom, transcending the knowledge. From this statement, Bello (2013) invites to reflexive transformation of the significances and manifestations of the human coexistence, leading to the creation of multi-diverse realities, the convergence and everyday transcendence at all levels.

In summary, the overview of this discipline entails a cybernetic approach of the continuous searching of the wisdom, in other words, scanning in the knowledge of the knowledge, favoring conditions that improve the human condition in order to ensure its survival. The human being in recognition of his own nature, does not obviate himself or his social coexistence, recognizes the urge to transcend.

The bioethics assessment, such as synergy of the Geoethics, obeys the imperative applicability not only at the organizational level, and specifically in the energy sector, but also in the general society, from an assertive environment of its own recognition and the otherness, in its own rights and duties, and also the communicative projection of an active orientation in the life.

The emergence of the Geoethics flows through an evolutive range of statements proposed by various scientists and academics with sensitivity to the environment and all forms of life on the planet; in conjunction with the concern about the exhaustion of the world's natural resources and the responsibility the humans have regarding the future generations, all these elements converge in a linked cosmovision of using today's technologies toward the societies.

According to Nikitina (2016), refers Geoethics from the ancient Greece with the planet's resources exploitation to an enslaved object in the service of humans. This perspective evolves epistemologically for the "illuminist" that in defense of private property perceives the territories, and moreover the space as elements with commercial value, that assure the humanity progress.

The same author states that toward the evolution of the Geology proceeds the recognition of the human being on Earth, with the introduction of actions that comes from the human intelligence, altering and modifying the planet layer where it influences (surface and subsoil); at the same time, proposals arise from proposals about the defense toward the preservation of nature.

According to the author, the transit of the Geoethics through the conservatism, "the ecology awareness," "the Eco-philosophy," "the ecologic ethics," and the regulations for the protection of the environment from various countries; the ethics of the Earth; and, the development of the model of "Ecocentric Ethics," the "Deep Ecology," the management of the common assets are predecessor of the Geoethics.

It is evidence, with this statements, that the path that leads to the Geoethics is marked by the recognition of the life, the environment, the ecologic postures respects to the preservation, and the recognition of the human being as a specie with awareness that gives it capabilities of constructing or destroying, in this case, the planet resources, which forms the environment of its daily development.

Nikitina (2016) refers to Vaclav Nemec as a pioneer of the ideas about the ethics development in the use of mineral resources, calling this trend: "Geoethics" inspired by the business Ethics and an idea of formulating a special ethics for geologists and miners. Likewise, the author projects this discipline as a way to do justice to the future generations, taking into account that the non-renewable resources belong to those generations, as well as the current population, and for both the developed and developing countries.

It is convenient to recognize the organizations as part of the society, and it refers to a group of persons with similar culture, relating to each other, following rules and standards of a legal and customary structure, in a space and fixed term; involving groups of humans with common interests (countries, professional societies, companies, social work workers, and others). Under these premises, the organizations are considered as systematic structures with intentions, where there exists an institutionalized mechanism which is tied to a codified conduct based on acceptable and required actions (values), according to a view of set purposes.

Therefore, there exists an institutional framework where the organizational ideology is very important because it forms its participants or members, through communication network about ideas, thoughts, and lines of organizational action. In this respect, Guedez (2004) proposes that companies, understanding them as a business, constitute a scenario where the ethical conduct is being tested, taking into account the economic decisions that involve.

The purposes of each action and its implementation are governed by the ethics; however, there exists gaps between the regulations and the facts outside of what is allowed by the social practices, therefore, it is referred to transgression and deformation of moral principles or institutional mission and involves destructive effects.

In this regard, the management modeling of the knowledge at the organizational levels is an element showing outstanding financial and organizational results, and also benefits in terms of cost for the companies. However, Nonaka and Takeuchi (1999) outline the key issues to synergistic expansion of the knowledge with the creation in conjunction of the person and organizational knowledge. Therefore, the creation of the theory of the organization knowledge is at the same time a basic theory for the construction of a truly humanistic knowledge society, beyond the economic or business perspective, where in turn, the communication between natural persons through the four ways of knowledge conversion is facilitated: socialization, combination, externalization, and internalization, process that is called: "spiral of knowledge."

It interacts in a dynamic and continuous context of an organization, while recognizing the role of the persons as essential actors in the creation of new knowledge.

The knowledge management in energy sector organizations promotes epistemically the knowledge production in an interactive field which tends to the innovation; promotes organizational structures where the new knowledge is accumulated, exploited, and created in a dynamic way to re-categorize and re-contextualize in a strategic manner to be used by other persons within the organization and future generations.

In order to relate the knowledge management in an axiological approach, at the energy sector organizations in Venezuela and Latin America, some objects of social-contextual dimensions included in the synchronous structure of the epistemic structure that comprises beliefs, standards and rules in personal and organizational dimensions, such as bioethics, ethics, and the ethics of the organization, values, codes of action, moral principles, among others are taking into account without displacing the corresponding complex thoughts, the current principles of the bioethics and the knowledge management.

In the entire history of humankind, the science passes through the true nature of new sources of energy that assures the power of domination, by means of using technologies and knowledge. In the field of energy, specifically in the hydrocarbon industry, the geologic knowledge contemplates the base of the knowledge which is the support of all potential reservoir management that allows for knowing the location, delineation and internal configuration, its genesis, and nature with more certainties.

This premise makes transcendental not only the development of hydrocarbon fields and basins, among other resources in an efficient manner, but also the decision-making based on the estimated volumes of the planetary resources (minerals, water, hydrocarbons, and others) for the construction of the economic model of each nation and region as a function of the State policies.

The relationship of the Geoethics and the knowledge management with the academic and energy sector organizations contribute epistemically to bioethics values and Geoethics processes in order to conduct assertive decisions from personal and group actions through the life taking into account the environment as a part of the man and himself as a complex representation of the surroundings, interrelated similarly toward the environment respectfully and the nature as a basis of the subsistence of the human species.

However, the Geoethics in the organizations of the academic and energy sectors is considered mainly unknown. There are not only employees of public and government companies and private sector (oil and electric energy companies), in several Latin American countries, but also academic employees who have been consulted with this respect in order to give testimonies and support this dissertation, and they are expressed their lack of knowledge about it. Performing the linkage of themes such as bioethics, axiological pertinence, and reflexive responsibility within the mentioned sectors in Venezuela is unprecedented making easier it starting point.

Therefore, this proposal deserves, as Bautista (2016) indicates, a path to follow that breaks with preexistent structures to generate a "persovision" and cosmovision of a different style, setting the epistemological bases of the investigative approach, an integrated way of the subject who investigates in conjunction with the studied object because any object or circumstance of the environment conforms part of the same system to which the human belongs.

In that sense, its events and synergies in aspects such as the Geoethics in the geologic knowledge management in energy sector organizations address an updated historic behavior about the use of our planet and the other elements of the universe, decanting in a reflection that drives to a reflexive pertinence innovating the energy and academic sector, through the insertion of this discipline in the epistemic geology from the academic, and also enhancing the training programs of the human potential of the companies related to the energy sector. This action motivates an ethical behavior of all participants/actors. The human potential as a tangible resource generator of value for companies will not be dissociated from the tangible value of the planetary resources in supporting the economic development of the society; instead they go together with the intrinsic value of the humanity, the life, and its preservation.

In such a way that, the picture in Fig. 14.1 dates to 1938 and shows the oil activity of the transnational companies when massively exploited the deposits of the Maracaibo Oil Basin. All this exploitation caused severe alterations of the ecosystem, modified the environment, filled the cities of the East Coast of Lake of Maracaibo with facilities called: “oil surface facilities,” among other damages. In addition, the largest freshwater reservoir in South America (Lake of Maracaibo) lost its original natural condition because it was transformed from an estuary, with connection to the Caribbean Sea through a navigation channel, to a body of brackish water whereby large oil and commercial ships navigate. This transformation was done under the pretext “of human purposes” for crude oil transportation to foreign ports.

Throughout the time of the oil exploitation, not only the environmental impact is latent, but also the risks to which the population that grew in the surroundings of these locations are exposed, since the subsidence generated by the imbalance caused by the extraction of the components of the subsoil has left populated areas below sea level, subject to overflowing by nature, the artificial protection of a perimeter wall containing the mentioned body of water. These correspond to topics of Geoethics attention.

Epistemologically, the approach of the complexity offers the possibility of taking into account distinct horizons concerning the requirements of the human dynamic in constant relationship with the planet, also, leads to several elements such as the use

Fig. 14.1 Oil facilities of the East Coast of Maracaibo Lake, Venezuela 1938 *Source* open web sites

of minerals as livelihood for developing the nations (covering the social needs), in addition to the hydrocarbon reserves, the fresh water and aquifers, in contrast with the incomes earned by its extraction, the use of this incomes by the governments according to public policies and international agreements within the external strategies, and the management of the resources from the transnational companies.

From the own Venezuelan's reality, within the Latin America context, proposals are generated to rethinking solutions for global problems based on the empowerment suggested by the huge natural resources within its territory, and it is therefore required to expand and disseminate the bioethics principles and Geoethics in all disciplines, not only in the episteme of the geology but also in several degrees of specificity, all the areas of knowledge.

Acknowledgements To the translator Eng. Roselba Canelon.

References

Bautista M (2012) Métodos Prospectivos. Compilación para el Programa del Doctorado en Ciencias mención: Gerencia de la Ciencia y la Tecnología. URBE, Venezuela

Bautista M (2016) Investigar en el continous de la complejidad. II Jornada de divulgación del programa de investigación y I encuentro de investigadores UNERMB. Venezuela

Bello R (2013) Gestión de la innovación en el contexto organizacional inteligente: visión epistémica desde el ámbito bioético/transcomplejo. Universidad Rafael Belloso Chacín: http://www.urbe.edu/portal-biblioteca/basesdedatos-urbe/tesis/. Tesis Doctoral. Consultado en 2015

Delgado C (2007) Hacia un nuevo saber. La bioética en la revolución contemporánea del saber. Editorial Acuario. La Habana

González JL, Martínez Frías J (2011) Geoethics: un reto para la deontología profesional. Tierra y Tecnología 39:62–66. Disponible en: http://www.colegiogeologosch.com.ar/archivos/biblioteca%20virtual/publicaciones/geoetica/Gonzalez%202011.pdf. Consultado: 20

Hayles N (1998) La Evolución del Caos. El orden dentro del desorden en las ciencias contemporáneas. 2da Edición. Editorial Gedisa. España

ICOG (2011a) Código Deontológico del Ilustre Colegio Oficial de Geólogos, http://www.icog.es/_portal/vu/codigo_deontologico_ICOG.pdf

ICOG (2011b) Ilustre colegio oficial de geólogos. Código deontológico. El Chubut. Disponible en: www.consejogeologia.org.ar/index.php/legal/codigo-de-etica. Consultado en 2017

Lenk H (2007) Global TechnoScience and responsibility. Lit Verlag, 416 pp. Consultado en 2016

Maldonado C, Gómez N (2010) El mundo de las ciencias de la complejidad. Un estado del arte. Documento de Investigación No. 76. Editorial Universidad del Rosario. Bogotá D.C

Martínez Frías J (2011) Geosciences: an open access journal on earth and planetary sciences and their interdisciplinary approaches. Geosciences (1):1–2 (Consultado: Abril 2016)

Maturana H, Varela F (1994) De Máquinas y Seres Vivos. Autopoiesis: La organización de los vivo. Editorial Lumen, Buenos Aires

Maturana H, Varela F (2011) El árbol del conocimiento. 19na Edición. Editorial Universitaria

Molina V (2008a) Ciencias sociales y complejidad. Revista Desacatos, núm. 28, septiembre-diciembre, pp 11–14. Centro de Investigaciones y Estudios Superiores en Antropologia Social-Distrito Federal, México. Consultado en 2017

Molina V (2008b) Ciencias sociales y complejidad. México

Morin E (1998) Introducción al pensamiento complejo. Gedisa, Barcelona

Morin E (2003). ¿Sociedad mundo, o Imperio mundo? Más allá de la globalización y el desarrollo Gazeta de Antropología Nº 19, Texto 19-01. CNRS, París
Morin E (2004) La epistemología de la complejidad. Gazeta de Antropología, Numero 20, artículo 02. Disponible en: http://www.ugr.es/~pwlac/G20_02Edgar_Morin.html. Consultado en 2017
Morin E (2011) La Vía para el futuro de la humanidad. Paidós, París
Navarro V Col. (1999) Algunas reflexiones de la bioética en las ciencias médicas. Revista del Ministerio de Salud Pública de la República de Cuba. Educ Med Super v.13 n.1 La Habana ene.-jun. Versión ISSN 0864-2141. Disponible en: http://scielo.sld.cu/scielo.php?script=sci_arttext&pid=S0864-21411999000100002. Consultado: 19/05/15
Nêmec V (2012) 2° World sustainability forum. Geoethics and sustainability. Sciforum. International Association for Geoethics (IAGETH). CZ-100 00 Praha 10 -Strasnice, K Rybnickum 17. Disponible en: https://sciforum.net/conference/wsf2/paper/900/download/pdf. Consultado en 2016
Nikitina N (2016) Geoethics: theory, principles, problems (segunda edn). (L. G. Monograph. M, Ed.) Moscu: Paidós. Recuperado el 09 de 2016, de http://www.icog.es/iageth/index.php/geoethics-theory-principlesproblems-nikitina-2016-second-edition-of-the-first-monography-worldwide-on-geoethic
Nonaka I (1994) A dynamic theory of organizational knowledge creation, vol 5. Institute of Business Research, Hitotsubashi University Organization ScmNcn, Tokyo, Kunitachi, Japan
Nonaka I, Takeuchi I (1999) La organización creadora del conocimiento. México, Oxford University Press
ONU sitio oficial. Disponible en: http://www.un.org/es/sections/what-we-do/promote-sustainable-development/index.html. Y Metas en cada objetivo: Disponible en: http://www.un.org/sustainabledevelopment/es/objetivos-de-desarrollo-sostenible/. Fecha de consulta: 20/09/16
Padrón J (1998) La estructura de los procesos de investigación. (R. e. Humanas ed) Revista educación y ciencias humanas. Año VII(15). Recuperado el 2016, pp 3–33
Padrón J (2016) Teoría de la Investigación. Una visión evolucionista y cognitiva de la producción de investigación. Universidad del Zulia, Venezuela
Potter V (1970) Bioethics: the science of survival. Perspectives in biology and medicine, Wisconsin University. Disponible en: http://www.pages.uoregon.edu/.../Potter_BioethicsTheScienceofSurvival.pdf
Potter V (1971) Bioética puente, bioética global y bioética profunda. Editado en University of Wisconsin-Madison, USA
Prigogine I, Stengers I (1984) Order out of chaos. Man's new dialogue with nature. Editorial Bantam Books, Canadá and EUA
Prigogine I (1997) El fin de las certidumbres. Editorial Taurus, Madrid
Rivas P (2017) Liderazgo bioético: una cosmovisión sistémica en las organizaciones del sector energético. Tesis Doctoral. Universidad Rafael Belloso Chacín URBE, Venezuela
Sibilia P (2006) El hombre postorgánico. Cuerpo, subjetividad y tecnologías digitales. Editorial del Fondo de Cultura Económica, Argentina
Szabó S (1997) Geoetika a jej princípy. Acta Montanistica Slovaca Rocník, número 2, 4, 347–350. Disponible en http://actamont.tuke.sk/pdf/1997/n4/6szabo.pdf. Consultado en 2016 y 2017
UNESCO sitio oficial. Disponible en: http://www.unesco.org/new/es/education/themes/leading-the-international-agenda/education-for-sustainable-development/sustainable-development/. Fecha de consulta: 20/11/16
Von Bertalanffy L (1986) Teoría general de los sistemas. Edición del Fondo de Cultura de México, México

The manufacturer's authorised representative in the EU is Springer Nature Customer Service Centre GmbH, Europaplatz 3, 69115 Heidelberg, Germany. If you have any concerns regarding our products, please contact ProductSafety@springernature.com

Printed and bound by CPI Group (UK) Ltd, Croydon, CR0 4YY
01/10/2026
02225860-0001